ECHIDNA
Extraordinary egg-laying mammal

ECHIDNA
Extraordinary egg-laying mammal

Michael Augee, Brett Gooden and Anne Musser

CSIRO
PUBLISHING

© Michael Augee, Brett Gooden and Anne Musser 2006

All rights reserved. Except under the conditions described in the Australian *Copyright Act 1968* and subsequent amendments, no part of this publication may be reproduced, stored in a retrieval system or transmitted in any form or by any means, electronic, mechanical, photocopying, recording, duplicating or otherwise, without the prior permission of the copyright owner. Contact **CSIRO** PUBLISHING for all permission requests.

National Library of Australia Cataloguing-in-Publication entry

Augee, Michael L.
 Echidna : extraordinary egg-laying mammal.

 Bibliography.
 Includes index.
 ISBN 0 643 09204 8.

 1. Tachyglossidae. I. Gooden, Brett (Brett A.), 1943– .
 II. Musser, Anne. III. Title. (Series : Australian
 natural history series).

599.29

Available from
CSIRO PUBLISHING
150 Oxford Street (PO Box 1139)
Collingwood VIC 3066
Australia

Telephone: +61 3 9662 7666
Local call: 1300 788 000 (Australia only)
Fax: +61 3 9662 7555
Email: publishing.sales@csiro.au
Web site: www.publish.csiro.au

Front cover
Photo by Chris Tzaros

Back cover
Photo by Gordon Grigg

Set in 10.5/14 Sabon
Cover and text design by James Kelly
Typeset by J&M Typesetting
Printed in Australia by Ligare

Contents

Preface — vii

1. The monotremes — 1
2. Evolution — 13
3. Skeletal anatomy — 33
4. The brain — 45
5. Senses — 55
6. Reproduction — 77
7. Behaviour — 89
8. Food and feeding — 95
9. Metabolism — 107
10. Conservation and management — 119

Glossary — 127

Bibliography — 133

Index — 135

Preface

This book is based on *Echidnas of Australia and New Guinea*, first published in 1993 as part of the Australian Natural History Series. We have considerably expanded the original work, incorporating recent research, most of which has been in the area of hibernation, palaeontology and field biology. Sadly, there has been almost no research into other aspects of physiology, although many questions remain to be answered. The increase in our understanding of echidnas in their natural environment is due in a large part to continuing field studies on Kangaroo Island being carried out by Peggy Rismiller.

Beyond doubt the major contributor to scientific knowledge of monotremes in the twentieth century was Mervyn Griffiths. His two books and many research papers are indeed the basis on which this book is built. His contribution, however, far exceeds the written word, and he inspired and guided students and colleagues without hesitation. His death in 2003 sadly preceded the third Symposium on Monotreme Biology, held at Sydney University in July of that year. That symposium – as well as this book – are dedicated to him in appreciation.

As in the original book, most of the photographs have been supplied by Professor Gordon Grigg, whose work, with that of his colleagues and students, has greatly expanded our understanding of echidna physiology. His firm belief that field studies are necessary to test theories based on laboratory studies was certainly validated by field studies of hibernation in echidnas, the results of which can be found in Chapter 9 of this book.

Additional photographs have been supplied by Barbara Smith, whose active support in field studies of echidnas is gratefully acknowledged.

M.L. Augee
B. Gooden
A. Musser

1
The monotremes

There are only three types of living monotreme – the short-beaked echidna, the long-beaked echidna and the platypus. They are mammals, and like all mammals they have fur and produce milk to nourish their developing young. But in a lot of ways they are not quite like their fellow mammals – the marsupials and the placentals. The most obvious difference is that platypuses and echidnas lay eggs (oviparity), and so their young are hatched, not born alive.

Since humans are placentals and humans write the textbooks, monotremes often get put in their place as 'almost' mammals or Prototheria in the Latin of scientific nomenclature. This placental-biased view of monotremes as some sort of early test model that wasn't quite right has tarnished them for two centuries. However, as we shall point out in this book, monotremes have been around for a lot longer than placentals and have remained masters of their environmental niches. That means that even if they are not very good at being humans or laboratory rats, they are in fact very, very good at being echidnas and platypuses.

Monotremes are often listed as being Australian, but both types of echidna are found in New Guinea, although the living long-beaked echidna is not found in Australia at all. The platypus is a 'fair dinkum' Aussie these days, but its ancestry includes a South American relative that lived in Argentina just after the dinosaurs disappeared from the scene.

The short-beaked echidna

The short-beaked echidna is a medium-sized mammal, rarely weighing more than seven kilograms, covered on its back and sides with stout spines amongst a fur coat of varying colour from light brown to black (Figure 1.1). Very lightly coloured individuals are occasionally reported as 'albinos', although, as can be seen in the photographs on page 26, while the eyes are pink there is pigment in the hair and spines.

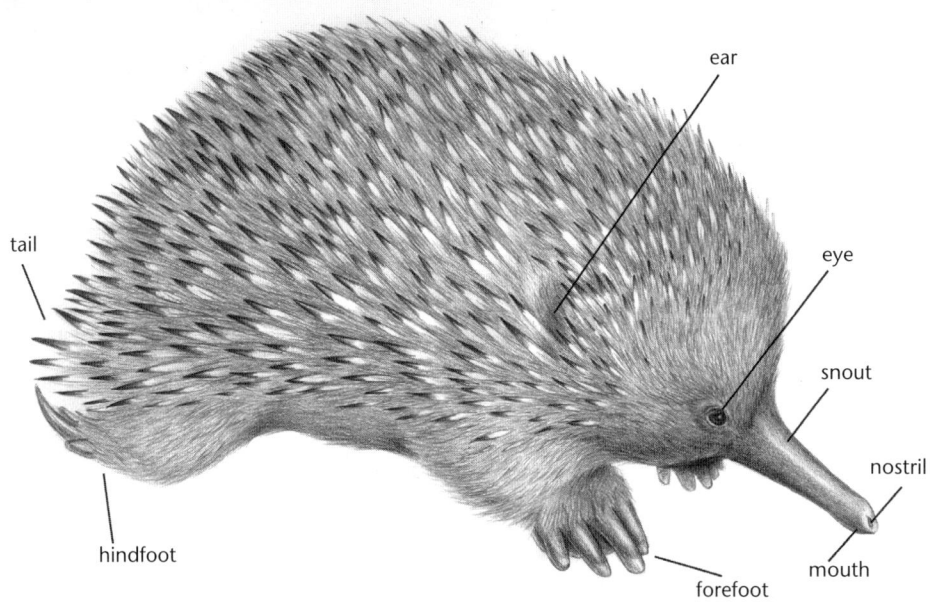

Figure 1.1 *Tachyglossus aculeatus*, a typical adult echidna from Tasmania.

The echidna's head appears small relative to its stocky body, particularly because it has no obvious neck. Its head tapers to a naked, cylindrical snout, which serves as a prod and lever in the search for the ants and termites that comprise the main part of its diet. The name 'spiny anteater' is therefore an apt description, but 'echidna' is much more widely used as the common name in Australia although that name came about rather by accident.

The eyes of the echidna are small, black and somewhat protruding – they are sometimes referred to, unkindly, as 'beady'. The ear opening is not readily visible – it is usually obscured by spines, but it is quite a large vertical slit. The external ear (the pinna) is unlike that of other mammals and is formed by a large cartilaginous funnel that is largely buried in a superficial muscle.

The limbs of echidnas are held horizontally away from the body, as they are in platypuses. The forelimb is short and stout, ending in a broad hand (manus). The manus has five distinct digits, each ending with a flattened claw, forming an effective spade for digging. The hindlimb is much less robust than the forelimb and the hindfoot (pes) is smaller and narrower than the manus. The bones of the lower hindlimb (the tibia and fibula) are rotated in echidnas so that the hindfeet are pointed to the rear. The claws on the hindfoot are long and thin compared with the spatulate claws on the forefoot. The second digit on the hindfoot is long and recurved and is often called the 'grooming claw'. Echidnas have a remarkable ability to groom in hard-to-reach places, even behind the neck, by rotating the hindlimb, twisting the foot and using the grooming claw to scratch between the spines. The tail is stubby; the spines form two handsome semicircular whorls over the tail area.

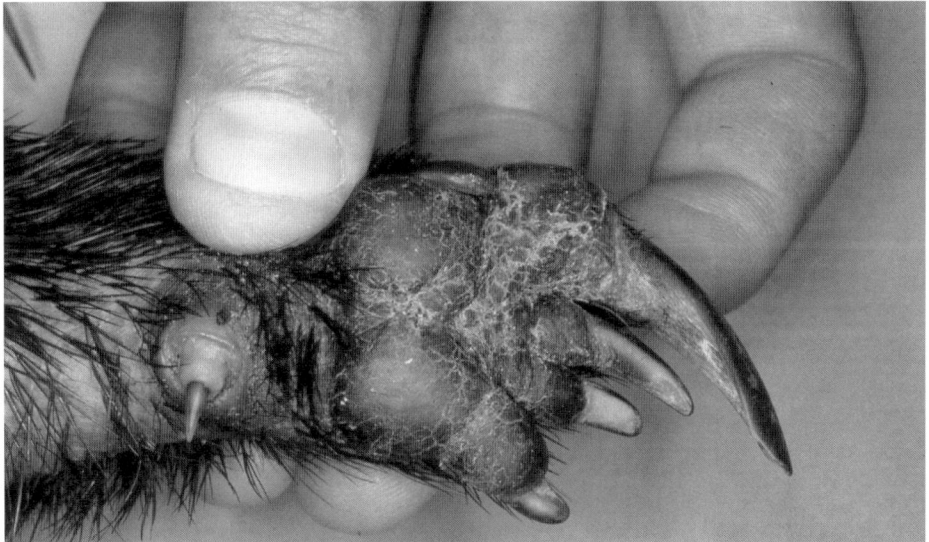

The hindfoot of a short-beaked echidna, showing its enlarged second or 'grooming' claw, as well as the crural spur in the region of its ankle.

Without picking them up (not an easy task!) for close inspection, it is impossible to distinguish male from female echidnas by their appearance. Although overall males are about 25 per cent larger than females, there is so much overlap that the sexes cannot be distinguished on size. There are no outward signs of sexual organs. Both sexes have only one opening leading from the cloaca to the outside, through which urine, faeces, and reproductive products must pass. The egg must pass through this opening when laid by the female and the penis protrudes through it in the male when mating. The penis of the male can usually be located by feel under the skin near the cloaca.

Adult male echidnas have a spur in the region of the ankle which is known as the crural spur. This is not a digit but an entirely separate structure. In the platypus the spur is connected by a duct to a venom gland lying behind the knee. In the echidna the spur is 0.5 to 1 cm in length and well developed. The duct and gland are vestigial. The spur may not be obvious until the fold of skin covering it is probed. The echidna does not appear to have the muscular ability to erect or retract the spur. In the young, spurs begin to develop in both sexes, but in the female the spurs regress and are not obvious. In the male the spur is covered by a sheath which is lost at some point before the animal reaches four years of age. Presence or absence of this sheath can be taken as an indicator of sexual maturity.

In our experience of many years working with echidnas, we have found in all cases where we were able to verify the sex of the animal by other means, such as determining the presence or absence of a penis, animals with unsheathed spurs were males. However, some authors have claimed that one or both spurs can be retained by adult females. Therefore the presence of a spur in an adult echidna can be taken as a useful (but not necessarily foolproof) indicator of male sex.

The pouch can likewise be an inconsistent indicator of sex. A depression on the ventral surface of the abdomen, bounded by two ridges of muscle, can be seen in both sexes. In pregnant females enlarged mammary glands form thick lips on either side of the midline and the resultant pocket envelops and protects the egg and, subsequently, the newborn young. At the front of the pouch area there are two small, hairy areas (known as areolar patches) where the milk glands open. Monotremes do not have teats or nipples and the young suck milk directly from the hairs over the openings of the milk glands. The pouch regresses after the young is independent. It is worth noting that the platypus does not have a pouch even though the milk delivery system is the same as it is in echidnas.

Discovery and early scientific studies by Europeans

The first written description of an echidna is found in Captain Bligh's log of the ship *Bounty*. The *Bounty* was on its way to Tahiti in 1792 and had stopped at Adventure Bay on Bruny Island off the coast of Tasmania for two weeks. The entry for 9 February 1792 reads:

> 'An animal shot at Adventure Bay. It had a Beak like a Duck – a thick brown coat of Hair, through which the points of numerous Quills of an Inch long projected these very sharp – It was 14 inches long & walked about on 2 legs. Has very small Eyes & five claws on each foot – Its mouth has a small opening at the end of the Bill & had a very small tongue. – W.B.'

The echidna was shot by George Tobin, a ship's officer, who reported: 'The animal was roasted and found of a delicate flavour'.

The scientific, as opposed to culinary, interest in echidnas began when they were brought to the attention of Western scientists shortly after European settlement of Australia. The first scientific name for the echidna was given by George Shaw in the *Naturalists Miscellany*, Volume 3 in 1792 as *Myrmecophaga aculeata*. By assigning this name, he placed the spiny anteater of Australia in the same genus as the placental anteater of South America. He illustrated his article with a painting made in Sydney by John White, a collector of plants and animals as well as a painter. White apparently gave an echidna to Governor Philip, who sent it to Joseph Banks by the ship HMS *Gorgon*. The specimen travelled in an eight-gallon cask with a number of marsupial specimens. They arrived in England in July 1792, and Shaw published his description by the end of that year.

In 1802 the British anatomist Everard Home recognised the relationship of the spiny anteater to the platypus, which he had earlier described as *Ornithorhynchus paradoxus*, and renamed the spiny anteater *Ornithorhynchus hystrix*. Generic-level differences between the platypus and the spiny anteater were duly noted, and the spiny anteater became *Echidna hystrix* shortly thereafter. *Echidna* refers to the Greek goddess Ekhidna who was half snake (reptile) and half woman (mammal), indicating that the possession of a mixture of reptile-like and truly mammalian characters by monotremes was recognised very early.

An important rule in designating scientific names is that a name given to one genus should never be used for another and that, when there is a conflict, the older use has priority. The name *Echidna* had been given to a genus of fish in 1788, and so in 1811 spiny anteaters were assigned the genus name *Tachyglossus* meaning 'rapid tongue'. However, 'echidna' lives on as the common name. The species name *aculeata* (meaning 'spiny') corrected to *aculeatus* (to grammatically fit with *Tachyglossus*) remains valid.

In the early days of European settlement of Australia, naturalists had a tendency to create many species based on colour, size and other characters that in due course were found to be highly variable. Later, as the concept of the biological species became accepted, a species was seen as encompassing all individuals with the capacity to interbreed. Consequently a number of earlier 'species' in the genus *Tachyglossus* were lumped into the single species *aculeatus*. Some of the older names have remained as subspecific designations, as listed in Table 1.1. We have not included *T.a. ineptus*, which was once applied to echidnas from Western Australia, since there are no characters of importance that separate it from *T.a. acanthion*. It is of course possible, even likely, that modern techniques of molecular biology, especially using DNA and MRNA analysis, will require a re-evaluation of these subspecies and perhaps even a splitting of the species *aculeatus*. In terms of morphology, the only consistent character that appears to divide the species is the state of the claw on the third digit of the hindfoot. In *T.a. aculeatus* and *T.a. setosus* it is as long as the grooming claw on the second digit.

Table 1.1. Subspecies of *Tachyglossus aculeatus*, the short-beaked echidna.

Name	Distribution	Distinguishing characters
T.a. acanthion	Northern Territory, northern Queensland, inland Australia and Western Australia	Hairs, usually black, are bristle-like, sparse on the back and often absent on the ventral surface. Spines long and stout.
T.a. aculeatus	eastern New South Wales and Victoria; southern Queensland	Spines overshadow fur, which is relatively short.
T.a. lawesii	New Guinea lowlands	Spines long and stout; fur thick and usually brown.
T.a. multiaculeatus	South Australia, especially Kangaroo Island	Many long, thin spines which project well beyond the fur.
T.a. setosus	Tasmania	Soft thick fur with spines relatively short and few. Spines rarely protrude above fur. Fur often light brown.

Distribution and habitat

Other than the house mouse, no other mammalian species can be found in so many divergent habitats. Short-beaked echidnas are found in the Snowy Mountains (where they hibernate over winter), in the tropics, in the tropical grasslands of the Northern Territory, throughout the arid zone and all along the coast. In fact a distribution map of echidnas includes all of the Australian mainland, Tasmania and other offshore islands (e.g. Kangaroo Island). Short-beaked echidnas also occur in New Guinea and on some islands off the main island (e.g. Salawati Island). On New Guinea *Tachyglossus* is found in lowland habitats, especially in eucalyptus woodland around Port Moresby, and at altitudes up to 1600 metres in the central highlands. However, *Tachyglossus* is rare in New Guinea today and, like the highland long-beaked echidna, *Zaglossus bruijni*, is a truly endangered species there.

In general, the subspecies of *Tachyglossus* listed in Table 1.1 can be related to habitat. The Tasmanian subspecies, *T.a. setosus*, occupies the southernmost part of the range – not surprisingly, it is distinguished by the thickness of its coat (which often obscures the spines). What is surprising is that on average Tasmanian echidnas weigh less than any other subspecies – about two-thirds the weight of mainland echidnas. This runs counter to Bergman's Rule, which predicts larger body size within a cline as the poles are approached. *T.a. acanthion*, which inhabits the hot, dry centre of Australia as well as some tropical zones, is almost hairless but weighs considerably more than *T.a. setosus*. The New Guinea and northeastern Queensland subspecies, *T.a. lawesii*, primarily inhabits tropical lowlands, confounding expectations by being much hairier than *T.a. acanthion* and heavier than *T.a. setosus*.

Longevity
Echidnas may enjoy exceptionally long life spans. One short-beaked echidna lived for nearly 50 years in captivity at the Philadelphia Zoo in the USA. Peggy Rismiller reports a free-living individual as having been observed over a period of 45 years. This relatively long life span is no doubt a benefit of life 'in the slow lane'.

The long-beaked echidna
The long-beaked echidna belongs to the genus *Zaglossus* (Figure 1.2). It is considerably larger than the short-beaked echidna and has a longer and more down-curved snout.

- The maximum weight recorded for *Zaglossus* is about 17 kg, compared to a maximum of about 7 kg for *Tachyglossus*.
- The snout of *Zaglossus* is about 10.5 cm long, compared to about 5.5 cm for *Tachyglossus*. There is considerable variation between indi-

Figure 1.2 *Zaglossus bruijni*, the long-beaked echidna of New Guinea.

vidual long-beaked echidnas in the degree to which the snout is bent downwards. *Zaglossus* also shows greater development of the tongue and salivary glands than does *Tachyglossus*.

Other differences include variation in the number of claws, and differences in their spines.

- While most *Zaglossus* have claws on all digits, many lack claws on the first and fifth digits of the hindfeet. *Tachyglossus* invariably has five clawed digits on the manus.
- The spines of *Zaglossus* are shorter and more solid than those of *Tachyglossus*. They also have a smaller lumen.

External features

Zaglossus usually has thick fur, varying from shades of light brown to black. Some individuals may have white markings on the face, paws and rump. Albino forms have been recorded.

There is a great deal of variation in the length and distribution of spines. Spines usually show only above the fur on the flanks, but Salawati Island forms are reported to have spines on the ventral surface (belly). Spines are usually light-coloured, although one type has been described as a separate subspecies (*Z. b. nigroaculeata*) because it has black spines. Other forms have black spines tipped with white, and some individuals have a mix of black and white spines.

As in *Tachyglossus*, *Zaglossus* males have a spur on the hindlimb and females usually do not. The spur has been described as 'well developed' in some males, however the scant information available suggests that the duct and venom gland are vestigial or absent.

It has been reported that *Zaglossus* develops a pouch for its young in the same way as *Tachyglossus* does, but this remains to be confirmed.

Long-beaked echidnas have been reported to make a soft snuffling, snorting sound.

Taxonomy

The long-beaked echidna was first brought to the attention of European scientists by the Dutch merchant and natural historian A.A. Bruijn. He had received a partial skull (lacking the lower jaw) of this animal from a native hunter from Mt Arfak in northern New Guinea. The skull was sent to Italy and subsequently described by Peters and Doria in 1876. They named this remarkable new echidna *Tachyglossus bruijni*, but several competing generic names were published in quick succession: *Zaglossus* by Gill in 1877, *Proechidna* by Gervais in 1877, *Acanthoglossus* also by Gervais, *Bruynia* by Dubois in 1881, and *Bruijnia* by Thomas in 1882. The fact that so many names were proposed for this singular animal highlights the interest generated by this unique monotreme as well as differences of opinion as to what constitutes a generic difference between two taxa. Anyway, the name proechidna was for many years the most well-known term for the New Guinea long-beaked echidna, although the generic name *Zaglossus* won out on the basis of priority. *Zaglossus* means great (za) tongue (glossus). The species name is on occasion given as *bruijnii*, but the reason for the double *ii* is unknown to us, and since it seems an unnecessary complication to a name that is already difficult, we will stick to the more widely used spelling 'bruijni' in this book.

The taxonomy of species and subspecies of *Zaglossus* has likewise been controversial. This is primarily because there is so much morphological variation within *Zaglossus*. Does this mean that there are several different species, or is this variation only a difference between races or subspecies?

The question of whether or not there are at least two separate species of *Zaglossus* goes back to shortly after its discovery when, in 1884, Dubois described a second species based on its small size, relatively straight beak and thick fur. A comparative study made by Allen in 1912 found no reason to separate *Zaglossus* into more than one species, and that view has dominated to the present. However, in 1998 Flannery and Groves proposed splitting *Z. bruijni* by resurrecting the species *bartoni*, which had been used by Thomas in 1907. The main character used to split *Zaglossus* is the number of claws on the forepaw, with *Z. bartoni* always having the full complement of five claws, while *Z. bruijni* always has less than five, with the first and fifth usually missing. The two species as resurrected by Flannery and Groves do not overlap in their geographical range at present. *Z. bruijni* occurs west of Lake Paniai, including the Vogelkop Peninsula and a small portion to the east in what is now Irian Jaya. It is also found on the island of Salawati, just to the north of the Volgelkop Peninsula. *Z. bartoni* is found only east of Lake Paniai, occupying the central highlands of most of the island of New Guinea. Flannery and Groves also erected a new species, *Z. attenboroughi*, on the basis of one skin and a crushed cranium. This is a weak basis on which to construct a new species, especially within a genus that shows so much variation, and we will not consider it further.

Distribution, habitat and diet

The habitat of *Zaglossus* is primarily highland forest (above 2000 m), although in some areas the distribution extends above the highland forest into alpine habitat and below it into hill forest. On Salawati, *Zaglossus* inhabits low-lying, flat, forested areas characterised by rich, deep soil.

Several species and genera of long-beaked echidna have been described from fossil deposits throughout mainland Australia and Tasmania. If these animals had the same habitat requirements as the living long-beaked species (wet forest with soft soil), it is likely the shrinking of their range and ultimate disappearance from Australia was due to the increasing aridity of the Pleistocene. *Zaglossus bruijni* has a fossil record in New Guinea (from the Nombe rock shelter site) dating back to the Pleistocene.

While echidnas in general are often termed 'spiny anteaters', long-beaked echidnas are mainly earthworm eaters. Their diet can also include varying amounts of small centipedes, scarab beetle larvae, lepidopteran larva and subterranean cicadas, depending on availability. The animal uses its long snout to probe the soft soil to locate prey, and its tongue has a unique adaptation for taking earthworms. The anterior third of the tongue has a deep groove on the upper surface which contains three longitudinal rows of backwardly directed, sharp, keratinous (like fingernail) spines. Earthworms are manoeuvred until they can be taken into this groove from either end. The tongue extends only 2–3 cm, which is much less than the 18 cm extension possible for the short-beaked echidna. On protrusion, the tongue of *Zaglossus* bends downwards, opening the groove in the process. When the tongue is retracted, the groove is tightly closed. This structure is not present in the short-beaked echidna which is a true anteater.

No echidnas have teeth, and both long- and short-beaked echidnas thoroughly masticate (mash) food items between horny plates at the back of their tongue and on the roof of their mouth (palate).

Food supply does not seem to be a limiting factor for long-beaked echidnas. Earthworms are abundant and relatively large in the humid montane forests inhabited by them. However, all *Zaglossus bruijni* are rare and are listed as 'endangered' by The World Conservation Union (IUCN). The decline in numbers is due to land clearing and the use of dogs and firearms in hunting. There has also been a breakdown of taboos against hunting that were in place before the spread of Western beliefs in New Guinea. Roasted in the coals of a fire, echidnas are considered a great delicacy by all peoples of New Guinea. Not surprisingly, *Zaglossus* is now absent from areas where human population densities are high.

The above information covers almost all that is known about the biology of long-beaked echidnas. The rest of this book will deal with the short-beaked echidna and for the sake of brevity the term 'echidna' can be assumed to refer to *Tachyglossus aculeatus* unless stated otherwise.

The platypus

Although this book is about echidnas, it is convenient to include a few details about the platypus for comparison.

Platypuses feed and mate in freshwater streams and lakes of eastern Australia, including Tasmania, King Island and Kangaroo Island (where they have been introduced). The platypus is predominantly an opportunistic predator on benthic invertebrates, mainly insect larvae, although small fish may also be taken. Like the echidna, the adult platypus lacks teeth. However, teeth are present in fossil platypuses and begin to develop in very young platypuses, only to be lost before they are fully formed.

The platypus takes food in through its open mouth and into cheek pouches where it sorts out the indigestible bits and spits them out. It then grinds the food items into a paste by means of hardened keratinous (horny) pads that have replaced the teeth.

Aquatic adaptations include webbed feet, a streamlined body shape and a waterproof and highly insulative coat. When underwater, platypuses close their eyes, which is therefore assumed to be an adaptation for swimming although other mammalian swimmers, such as otters, do not close their eyes in the water. When underwater, platypuses also close the flaps of skin over their nostrils, effectively shutting down the sense of smell through the nasal passages. This raises the question of how platypuses navigate underwater, find their prey and keep from bumping into things. It has been suggested that electro-receptors located in the snout of the platypus (and also present in echidnas) might serve these functions, but they are passive detectors of electric currents and it is unclear how they could respond to anything but large prey items such as yabbies.

Male platypuses, like male echidnas, have a sharp spur on the inside of the ankle. This interesting structure is not found in any other living mammals, although some fossil Mesozoic mammals may have had such structures. In the male platypus venom is produced by a gland lying behind the knee and delivered through a duct to the spur. Though functional, the spur is of questionable use. It is not used in capturing prey, which are relatively small invertebrates and fish which can be taken into the mouth whole. The fact that the spur does not develop in females suggests some sort of specialised sexual function, but its use in sexual encounters has never been observed and it would certainly be a matter of 'overkill' since the venom is strong enough to kill a female platypus. Male platypuses often have wounds apparently caused by spurs, but male to male combat using spurs has not been observed.

The platypus is a burrower as well as a swimmer – it digs complex burrows in the banks of streams. Nesting burrows, much larger than camping burrows, end in a chamber lined with leaves and grass for newly hatched young. For an animal with webbed feet to also be an efficient digger requires special adaptations. Digging is achieved by folding back the webbing on the forepaws to free the claws for the task of excavation. In swimming there is some rotation of the tibia and fibula to turn the hindfeet towards the rear, but in normal stance the hindfeet of the platypus are not pointed backwards as they are in echidnas.

Relationships

The relation between the living monotremes and between monotremes and the other major mammalian groups is reflected in the traditional divisions of the living mammals:

Mammalia				
Prototheria			Theria	
Monotremata (monotremes)			Metatheria (marsupials)	Eutheria (placentals)
Ornithorhynchidae	Tachyglossidae			
Ornithorhynchus platypus	*Tachyglossus* short-beaked echidna	*Zaglossus* long-beaked echidna		

There have been several recent attempts to revise the traditional taxonomy, including a proposal to put platypuses and echidnas in separate orders, but in our opinion these are not well supported by available data and merely serve to create confusion by redefining established names.

2
Evolution

The first mammals evolved during the Mesozoic era, over 200 million years ago. These early mammals lived alongside the dinosaurs, pterosaurs and giant, long-extinct marine reptiles. Monotremes, with their many ancient anatomical and physiological features, have traditionally been thought of as Australasian survivors of this early mammalian radiation. Many of the earliest mammals and the last of the mammal-like reptiles, the therapsids, had anatomical features similar to those in monotremes. Monotremes seem to have their closest similarities with these extinct animals, rather than with other living mammals.

Several types of fossil Mesozoic monotremes are now known. The material by which they are identified consists primarily of lower jaws, some of which retain teeth and some of which are toothless. This once diverse group has been reduced to just two families today: the platypus (Ornithorhynchidae) and the echidnas (Tachyglossidae). These two quite different types are the last two branches on what may have been a very long evolutionary tree. Recent fossil material is providing strong evidence for a very distant origin of the monotremes. It also supports traditional comparative anatomical studies which have long suggested that they arose at the base of the mammalian tree.

The oldest monotreme fossils are recorded from the Early Cretaceous period when Australia was the far eastern terminus of the supercontinent of

Gondwana. Australia would have been unrecognisable as the continent we know today. It lay far south, joined across its south-eastern coastline to Antarctica and via Antarctica to South America (comprising the region known as East Gondwana, centred over the South Pole). This polar environment nurtured the early monotremes as well as other types of small, archaic mammals, dinosaurs, marine reptiles such as the massive pliosaur *Kronosaurus*, flying pterosaurs and more 'modern' animals such as turtles, crocodiles and lungfish.

Birds also make an appearance in the Australian fossil record in the early Cretaceous. Australia's polar position gave it a cool, temperate climate with long winter nights and summer days. Although the winters could be extremely cold in the region that was to become south-eastern Australia, it was not ice-bound, as the Antarctic region is today. Australia was then covered in forests of ancient conifers, ferns and cycads. There are no modern environmental equivalents – perhaps the best analogy would be to modern day Tasmania without the flowering plants (angiosperms) that did not dominate environments until much later in the Cretaceous.

The oldest monotreme to date, *Teinolophos trusleri*, is known from just a few lower jaws found by Tom and Pat Rich and their colleagues at the Flat Rocks locality along the Victorian coast. The Flat Rocks site dates from the Early Cretaceous and is about 115 million years old. *T. trusleri* was a tiny, mouse-sized monotreme that was neither platypus nor echidna but something very different. The Flat Rocks locality has also produced dinosaur fossils, so this miniature monotreme had a much more precarious existence than platypuses and echidnas do today. This part of Victoria may have been quite cold at this time – there is possible evidence of permafrost in rock outcrops just metres from the spot where *T. trusleri* was found. Another group of interesting, enigmatic small mammals (the ausktribosphenids) have also been found at Flat Rocks, but their relationships are still in question.

The next-oldest described monotreme is the 110-million-year-old *Steropodon galmani* from the Lightning Ridge opal fields of New South Wales. This was the first Mesozoic mammal to be found on any of the Gondwanan continents and was a much larger monotreme (about the size of a small platypus today) than *T. trusleri*. *Steropodon galmani* is quite a spectacular fossil, because the original bone of the animal has been replaced with silica in the form of coloured, translucent opal. Although *S. galmani* is known only from a partial jaw with three molar teeth, the shape of its teeth are distinctive and can be firmly linked to those of more recent fossil platypuses. The Lightning Ridge site, slightly younger and more northerly than Flat Rocks, does not seem to have been as cold as the Victorian site and some of the animals found at the Ridge are not quite as archaic as they were further south. There are several additional monotreme-like jaws from Lightning Ridge under study, but all of them are incomplete and teeth are not present. They do show that there was a range of sizes and forms of monotreme mammals living together at Lightning Ridge and that they were probably the dominant mammal group in this area at that time.

Even small jaw fragments can provide evidence about the pattern of evolution. The miniscule jaw of *Teinolophos trusleri*, for example, provides significant evidence about the evolution of monotremes. Behind and below the tooth row, on the inside of the jaw, *T. trusleri* had a long, curved groove that held several jaw bones in addition to the dentary. The dentary is the only jaw bone in the lower jaw of living mammals and their ancestors whereas reptiles (including therapsids) and near-mammals had a series of bones in their jaw behind the dentary (Figure 2.1). These postdentary bones were progressively lost in the line or lines leading to modern mammals. *T. trusleri* probably retained the angular and pre-articular bones, which were reduced and transformed into part of the middle ear apparatus in the ancestors of mammals. This remarkable discovery is compelling evidence that early monotremes and their ancestors had an almost reptilian, rather than mammalian, middle ear and that monotremes may have independently evolved the type of middle ear seen in marsupials, placentals and their ancestors.

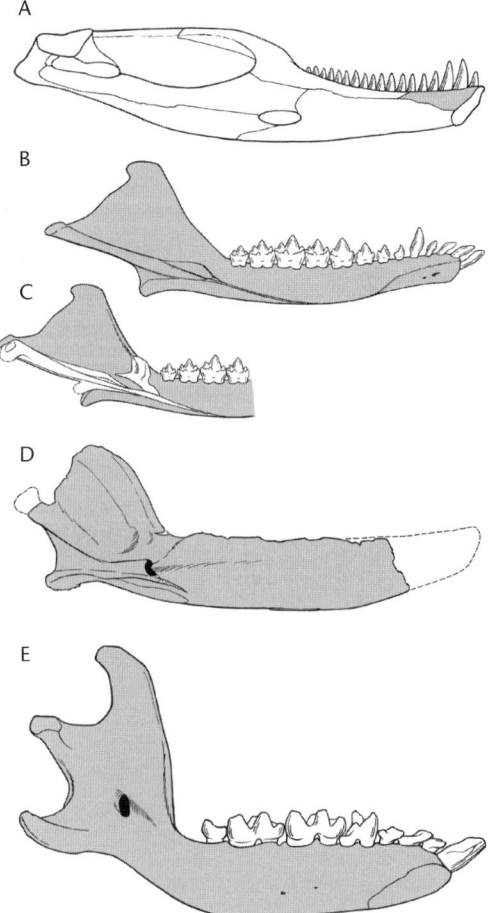

Figure 2.1 The position and size of the dentary (shaded) bone in the mandible of (A) the Permian reptile *Labidosaurus* from Romer 1956, (B) *Morganucodon* (postdentary bones omitted) from Kermack et al. 1973, (C) *Morganucodon* (postdentary bones reconstructed) from Kermack et al. 1973, (D) *Teinolophus* from Rich et al. unpublished, and (E) *Erinaceus*, the extant European hedgehog.

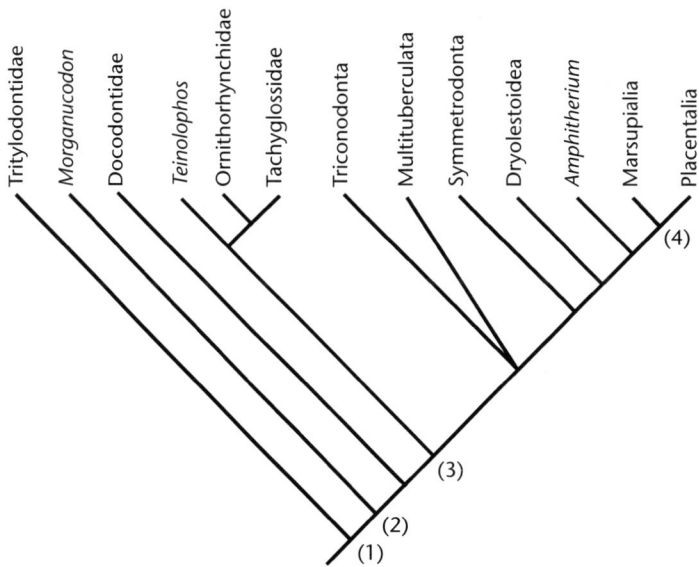

Figure 2.2 A cladogram showing the relationship between monotremes and various therapsid and mammalian lineages. Based on Rich et al. 2005. (1) Cynodontia, (2) Mammaliaformes, (3) Mammalia, and (4) Theria.

This discovery adds to the weight of anatomical and palaeontologic evidence placing monotremes in a very basal position within Mammalia, giving the echidna and platypus an exceptionally long pedigree (Figure 2.2). The *T. trusleri* jaw and the monotreme shoulder girdle are at a similar evolutionary level and, along with other archaic features, link monotremes to the earliest mammals or near-mammals rather than to Theria, the group to which marsupial and placental mammals belong. Monotremes may thus be by far the oldest mammals known, perhaps close to 200 million years of age – an extraordinary record of longevity for these strange and wondrous beasts.

Echidna evolution: variations on a theme

Echidnas suddenly appear in the fossil record about 15 million years ago, looking very much like echidnas of today – they were toothless, had long snouts and strong limbs. There are no earlier fossil monotremes that could conceivably be the direct ancestors of echidnas – no toothed monotremes with long, narrow snouts, robust limbs or other tachyglossid features. There is a gap of 50 million years between the first fossil platypus and the first fossil echidna. In all probability, echidnas are therefore the last of the monotremes to originate and diversify, and are thus the most recent (and likely to be the last) branch of this long and distinguished lineage. Because fossil echidnas are already toothless, following a dental trail is not possible. How did echidnas lose their teeth, and when? What were their ancestors like, and could we recognise them in the fossil record if they do appear? These questions can only be answered as new fossil material surfaces and potential echidna forebears make an appearance.

Fossil species

As with living echidnas, there are basically two types of fossil echidnas: large, long-beaked forms (several species from the Miocene through Pleistocene) and the short-beaked echidna *Tachyglossus aculeatus* (first recorded from the comparatively recent Pleistocene).

Long-beaked fossil echidnas have several features, as well as a longer fossil record, suggesting that they are more archaic than the short-beaked echidna, a relatively late arrival. Fossil long-beaked echidnas have been found in nearly every state of Australia as well as in New Guinea. However, the bulk of the fossil material is from south-eastern Australia: southern Queensland, New South Wales, Victoria, South Australia and Tasmania. Many of these fossils have been found in caves. Perhaps this is because echidnas in the past, like those today, used caves for shelter and thermoregulation.

The oldest echidna known, *'Zaglossus' robusta*, was discovered in 1895 at the bottom of a goldmining shaft in the New South Wales town of Gulgong. The fossil material consists of a large skull (with the end of the snout missing) and a humerus (upper arm bone), which was originally described as that of an enormous platypus. This material, although fragmentary, shows us that *'Z.' robusta* was a large echidna similar in overall size but more robust than New Guinea *Zaglossus*, with a long, toothless snout. Although the site was originally thought to be Pliocene in age, subsequent dating of the basalt overlying the site suggests that this material is from the much earlier Miocene epoch (13–14 million years). However, preservation of the fossil material is very similar to that of younger Pleistocene fossils from nearby Wellington Caves, which are in deposits that are less than one million years old although they lie within caves in limestone that dates from the Devonian (450 million years ago).

It is impossible therefore to rule out the possibility that the fossil-bearing material at Gulgong was deposited under the basalt long after the basalt had been laid down. If the earlier date were valid, *'Z.' robusta* would be much older than any other known echidna, none of which is older than the Pliocene (about four million years). The mineshaft has now collapsed and further excavations are unfortunately not possible. Both the date and the status of this animal must therefore remain in doubt. (Putting the genus name in quotation marks indicates the possibility that a specimen might belong to a genus other than the one to which it is currently assigned.)

Pliocene and Pleistocene long-beaked echidnas are currently assigned to two genera: *Megalibgwilia* (which includes most of the extinct Australian forms) and *Zaglossus* (northern Australian forms that are very similar to the living New Guinea *Zaglossus*). This division is debatable, and it is possible that when more analyses are complete most of these fossils will be recombined into the genus *Zaglossus*. All of these long-beaked echidnas were large and robustly built, with long, variably curved snouts. The shape of the snouts in these fossil forms, similar in general form to that of the living *Zaglossus* although not as strongly recurved, suggests that the diets of these echidnas may also have been primarily insect larvae and earthworms, rather than ants or termites. Insect larvae and earthworms would have been abundant in forest environments or grasslands, habitats that were lost in areas of Australia that are now semi-arid. It is interesting to note that the only surviving long-beaked echidna lives in relict forest areas of New Guinea. This type of environment was once much more common in Australia than it is today, where only pockets of rainforest remain.

One mainland long-beaked echidna deserves special mention: *'Zaglossus' hacketti* from the Pleistocene of Western Australia (Figure 2.3). This poorly known, enormous echidna was about one metre in length and is the largest echidna known. The relatively long length of its leg bones suggests that the

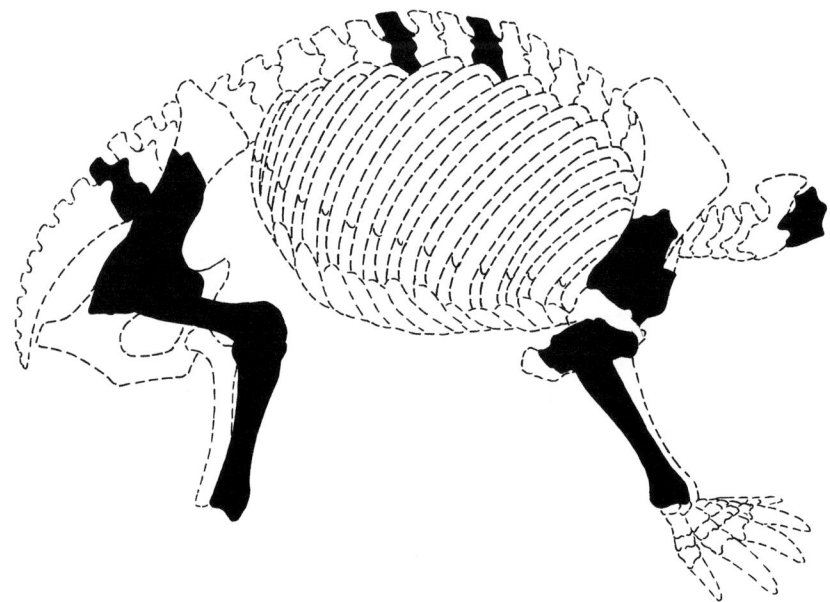

Figure 2.3 *Zaglossus hacketti* **from the Pleistocene of Western Australia, a poorly known, enormous echidna about one metre in length, making it the largest echidna known. Shaded bones are fossil remains, the rest of the skeleton is reconstruction.** Redrawn from Murray 1984.

	PERIOD	EPOCH	MONOTREME FOSSIL RECORD
CAINOZOIC	Quaternary	Holocene	
		Pleistocene	Ornithorhynchus anatinus Megalibgwilia ramsayi 'Zaglossus' hacketti Zaglossus species Tachyglossus aculeatus
	Tertiary	Pliocene	
		Miocene	'Zaglossus' robustus Obdurodon dicksoni
		Oligocene	Obdurodon insignis Obdurodon sp. A
		Eocene	
		Paleocene	Monotrematum sudamericanum
MESOZOIC	Cretaceous	Late	
		Early	Steropodon galmani Teinolophos trusleri
	Jurassic	Late	
		Middle	
		Early	evolutionary grade of ancestral monotremes as indicated by jaw form, shoulder girdle
	Triassic	Late	
		Middle	
		Early	

PALAEOZOIC

Figure 2.4 The occurrence of monotreme fossils through time (from Musser 2003). Names are listed by their first appearance in the fossil record.

posture and locomotion of this echidna differed from that of the other large forms. Perhaps it was able to navigate through heavily forested areas strewn with large logs and litter. Unfortunately the skull of 'Z.' *hacketti*, necessary for taxonomic identification, isn't known, and it is therefore uncertain if it is a '*Zaglossus*' or deserves to be given a separate genus name.

The short-beaked echidna, *Tachyglossus aculeatus*, makes its appearance in the Pleistocene (Figure 2.4). Its fossil record comes primarily from cave deposits in South Australia, where it has been found in some of the same deposits as Pleistocene long-beaked echidnas. Pleistocene *T. aculeatus* is indistinguishable from living *T. aculeatus* except for its size: the living form is about 10 per cent smaller than some of the fossil forms. This phenomenon, known as 'post-Pleistocene dwarfing', is seen in many Australian mammals known from the Pleistocene to the present (e.g. grey kangaroos and Tasmanian devils) as well as mammals from other parts of the world.

On the basis of the fossil record, echidnas do not appear to be as ancient as platypuses – they did not waddle past the last of the dinosaurs, as the platypuses did. On the other hand, molecular dating has projected the echidna/platypus split back to near the boundary between the Cretaceous and the Tertiary (about 65 million years ago). In either case, echidnas belong to the most ancient mammalian group known, now reduced to just themselves and the platypus. Although echidnas were never a diverse group, there were several species over the course of their evolution that found niches as insectivores or anteaters in a changing Australian landscape. The Pliocene-Pleistocene echidnas represent the last radiation of monotreme mammals. That radiation began to decline as long-beaked echidnas disappeared from mainland Australia when the continent became more arid in the Quaternary Period and the wetter environments that favoured the preferred prey of insect larvae retreated to the coastal areas. The occurrence of *Zaglossus* in New Guinea represents the 'last stand' for this archaic type of echidna, surely one more reason to protect its habitat into the future.

Anteaters – a case of convergence

Short-beaked echidnas are not the only mammals that have evolved to take advantage of social insects (ants and termites) as their primary food source. There are placental and marsupial anteaters as well (Figure 2.5). Anteaters have reduced or absent teeth; a long, thin tongue coated with sticky saliva; and a long, narrow, rounded snout for probing into areas where prey might be. They have powerful forelimbs, forepaws and claws for digging and for breaking into tree roots and trunks. Most anteaters depend on passive defence and have evolved structural defences, in the form of sharp spines or thick skin formed into interlocking or overlapping plates, or behavioural defences, such as burrowing or burying themselves, or foraging at night. Although anteaters from all three mammalian lineages have long, separate evolutionary paths, they have evolved similar structural and behavioural strategies to capture and eat their prey.

Figure 2.5 Silhouettes of anteaters (not to scale). From top: giant anteater (Edentata), armadillo (Edentata), numbat (Marsupialia) and pangolin (Pholidota).

Several unrelated placental groups have evolved ant-eating types. In South America there are three anteaters (giant, lesser and dwarf) in the family Myrmecophagidae as well as armadillos in the family Dasypodidae (which also has a North American representative). South American anteaters are completely toothless and armadillos have a very reduced dentition. Pangolins are armoured anteaters found in Africa and Asia. They are not only toothless but also lack the muscles required for chewing.

While marsupials are also found in the Americas, specialised marsupial anteaters have evolved only in Australia. The numbat (*Myrmecobius fasciatus*) is a termite specialist, although it has a full set of teeth which adults do not appear to use even in chewing. Marsupial moles (*Notoryctes* sp.) and striped possums (*Dactylopsila trivirgata*) are also anteaters.

Monotremes – the 'primitive' problem

The three major groups of living mammals – monotremes (Prototheria), marsupials (Metatheria) and placentals (Eutheria) – share the basic, defining characters of mammals including lactation (mammary glands), fur, and the ability to produce body heat from internal metabolic sources (endothermy). Marsupials and placentals share a number of characters not found in monotremes, such as live birth. Therefore marsupials and placentals are placed in the taxon Theria which excludes monotremes.

Table 2.1. Summary of monotreme characters
Derived (apomorphies)
    stocky, neckless bodies
    short, stout limbs held horizontally from body
    snout formed by elongated nasal bones and mandibles
    snout covered with skin rich in mechanoreceptors
    electro-receptors in snout
    little development of external ear pinna
    poison gland and spur on hind leg
    hindfeet rotated outwards
    replacement of teeth with keratinised pads

Ancestral (plesiomorphies)
    oviparity (egg-laying)
    large septo-maxilla (a wedge-shaped bone in the snout)
    post-temporal opening
    interclavicle and precoracoid in shoulder girdle

Monotremes possess several features retained from ancestral forms which have been lost in the common ancestor of the therian lineage. Monotremes, or at least platypuses, also have a very old fossil record. For these reasons monotremes have often been tagged 'primitive' mammals. As will be demonstrated clearly throughout this book, 'primitive' does not mean inferior, and living monotremes are highly specialised for, and successful in, their different niches. The term 'plesiomorphic' is better used for ancestral characters, as this more neutral term does not imply any condition of function or adaptation. Table 2.1 lists some of these plesiomorphic characters, as well as characters that are restricted to platypuses and echidnas and are assumed to have evolved only within the monotreme lineage (known as derived or apomorphic characters). These characters are discussed further in the relevant sections of this book.

The advantages of being 'primitive'

Some of the attributes of echidnas that have been labelled 'primitive' may in fact be the key to the incredibly long survival of monotremes. As we shall see later, there are a number of physiological and behavioural features of living monotremes, some of which have been considered 'primitive', that may in fact have been important factors in their long-term survival. For example:

- As a burrower, echidnas can avoid most predators and extremes of temperatures, both hot and cold.
- Body temperature can be lowered and controlled in the face of extreme cold or food deprivation.
- The food sources of echidnas are all to be found underground, not above it. Echidnas can forage in the dark, even surviving total blindness, by using other highly attuned senses such as electroreception.
- Echidnas are exceptionally tolerant (for mammals) to asphyxia, both acute, such as to airway blockage, or chronic, such as in response to low oxygen and high carbon dioxide in the available air supply.

With such faculties echidnas today are well equipped to survive minor crises such as bushfires, floods and cold winters. Perhaps such 'primitive' faculties enabled monotremes to survive longer and more challenging environmental changes that led to the extinction of other, more 'advanced' mammalian types.

The short-beaked echidna, *Tachyglossus aculeatus,* is found throughout the Australian mainland, Tasmania and other offshore islands, and in New Guinea. Photo: Chris Tzaros.

Some echidnas are unusually light. They are often referred to as 'albinos' but they are clearly pigmented. Photo: Gordon Grigg.

The snout, with its array of sensors, is used as a probe in search of food. Photo: Barbara Smith.

Echidnas have a remarkable ability to groom in hard-to-reach places by rotating the hind limb, twisting the foot and using their grooming claw. Photo: Barbara Smith.

The echidna's snout is often moistened with saliva. Perhaps this is an adaptation for making good electrical contact with the soil. Photo: Barbara Smith.

With its powerful claws, an echidna tears apart a log in search of food. Photo: Gordon Grigg.

A cardboard box is not a good container for an echidna. In a matter of seconds an unwatched echidna is up and away. Photo: Gordon Grigg.

Echidna scats are usually smooth and cylindrical with broken ends. They consist of insect cases in dirt. Photo: Gordon Grigg.

A male follows a female during the mating season. In the wild, trains of up to 11 echidnas have been observed. Photo: Gordon Grigg.

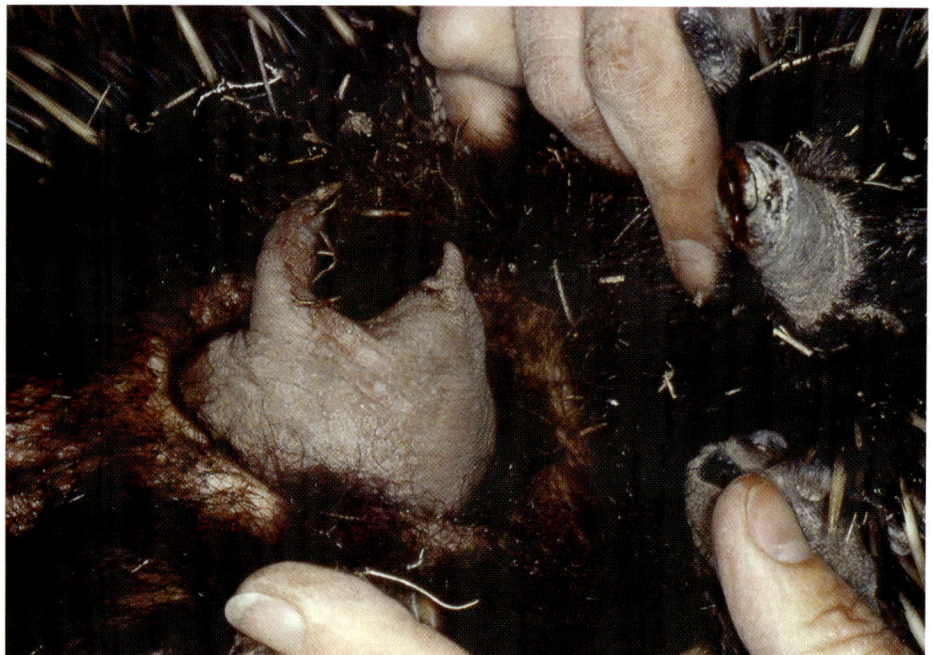

A young echidna, about 40 days old, can no longer fit entirely in its mother's pouch.
Photo: Gordon Grigg.

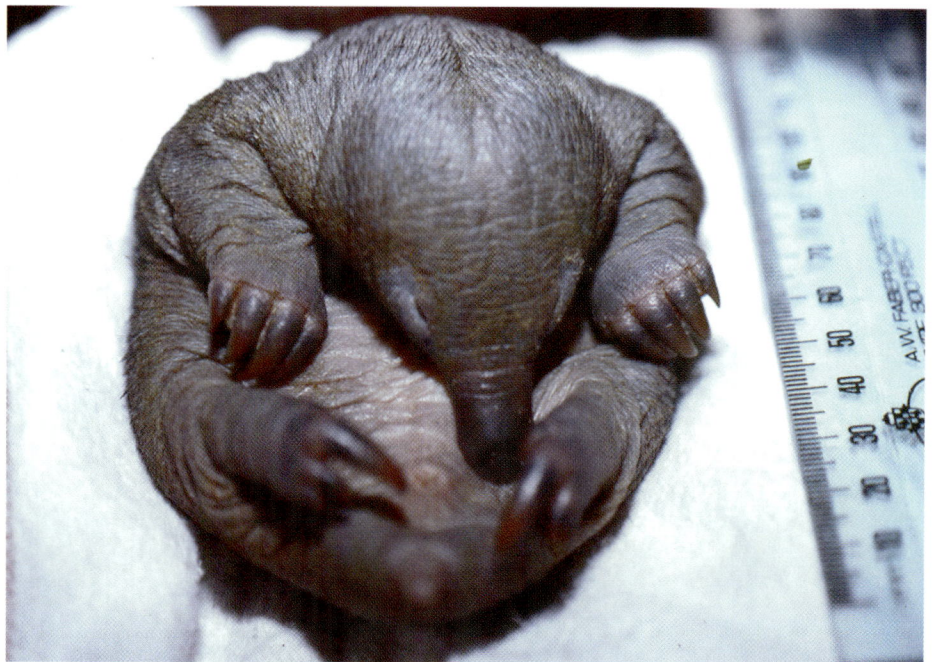

At 56 days old, this echidna weighs about 250 g.   Photo: Gordon Grigg.

This nursery burrow was excavated after the mother ceased to return. It had three chambers, less than 30–40 cm underground. Photo: Gordon Grigg.

This echidna has become active in the snow in spring in Australia's southern alps.
Photo: Gordon Grigg.

Echidnas have been observed swimming in dams in hot weather and swimming across streams. Photo: Barbara Smith.

In the wild, an echidna may occasionally drink from open water. Photo: Chris Tzaros.

3
Skeletal anatomy

Details of echidna skeletal anatomy are scattered throughout the literature, particularly in literature of the nineteenth and early twentieth centuries, often written in German. In this chapter we will attempt to bring together these historical but fundamental anatomical studies along with recent anatomical research. In the process, emphasis will be placed on the relationship between anatomical features of monotremes and those of evolutionarily significant groups, such as ancestral therapsids, the earliest mammal groups, archaic therian mammals (from whom monotremes are assumed to have diverged early) and the relatively advanced therians (marsupials and placentals). The significance of these comparisons can be appreciated with reference to Figure 2.2.

More specific details of anatomy are covered in those chapters that relate specific structures to specific functions. The digestive tract for example is described in Chapter 8 and reproductive structures in Chapter 6.

Bill, snout, beak or nose?

Platypuses are always said to have a 'bill'. The word 'bill' is used by analogy with ducks – indeed the platypus is sometimes known as the 'duck-billed platypus'. Carrying the bird analogy further, the two echidnas are usually referred to as 'short- and long-beaked'. The bones that form the top part of that structure are much the same in platypuses and echidnas (Figure 3.1), including a pincer-like shape at the front end. In platypuses the structure is flattened compared to the cylindrical shape of the echidnas. The surface is not similar to the hard, finger-nail like bird beak. In monotremes these structures are covered by a shiny, black, hairless skin that is rather soft and slightly moist to the touch. The skin contains many sensory receptors. This structure is more like the snout of a pig, for example, and like the pig snout, that of platypuses and echidnas includes the lower jaw and the opening of the mouth (at the very tip in echidnas). A snout is also a more appropriate comparison than a nose, which is a cartilage-supported structure that contains the nostrils only. For reasons quite unclear, the common names 'short-beaked' and 'long-beaked' for

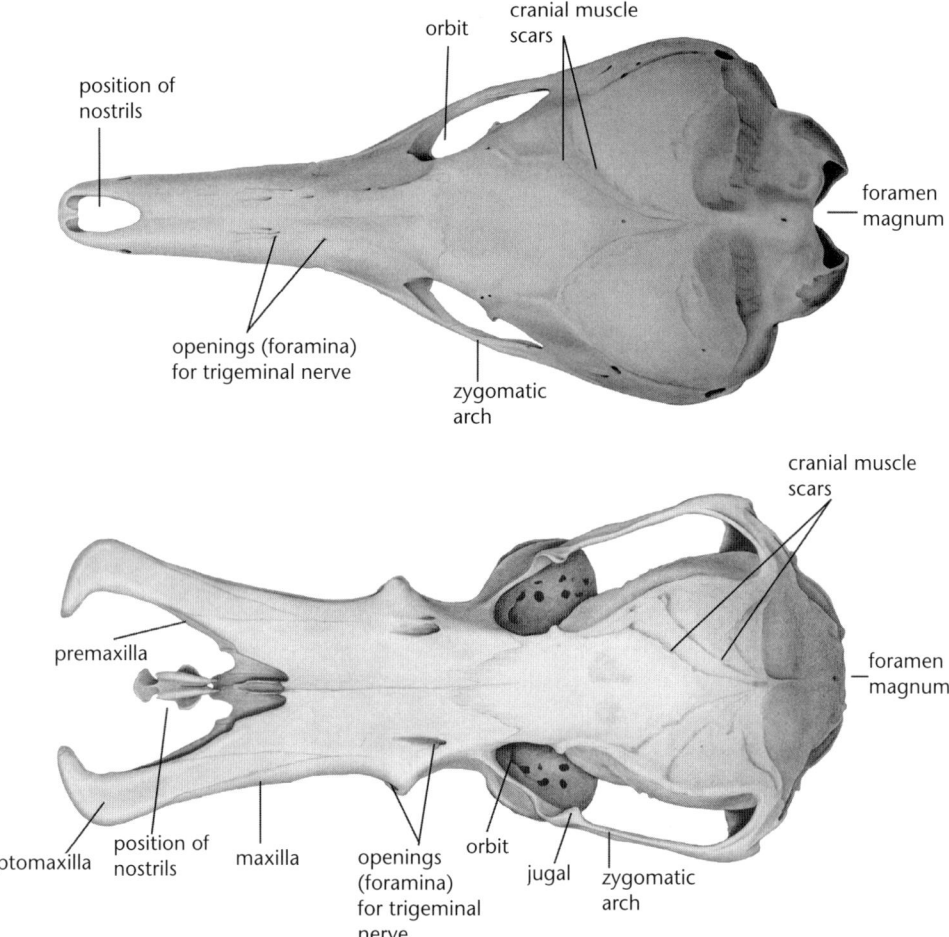

Figure 3.1 Dorsal view of the skull of the echidna (top) and platypus (below). The sutures on the snout are well fused on this echidna skull, and sutures between the premaxilla, septomaxilla and maxilla are not visible. Echidnas lack jugals although platypuses possess these small bones.

echidnas have become established and are the common names recommended for *Tachyglossus* and *Zaglossus* respectively by the Australian Mammal Society. We will follow this convention for the common names, but we will refer to the anatomical structure itself as the 'snout'.

Spines or quills?

The fur coat (pelage) of echidnas displays a gradation from fine hairs to sharp, keratinised spines. All echidnas have spines, although they are much more numerous and obvious in the short-beaked echidna than they are in the long-beaked echidna. Sometimes they are referred to as quills, a term which is used for similarly modified hairs in porcupines. 'Spine' of course is also a name used for the vertebral column. However, when referring to pelage, spine and quill are used interchangeably. There is a general conception that quills are relatively long and thin, as in the quill of a feather, compared to spines. Therefore the term 'spine' is appropriate for the modified hairs of echidnas while 'quill' is more appropriate to porcupines (a group of rodents not found in Australia) because they are much longer, thinner and more delicate than those of echidnas.

The spines on the stubby echidna tail form two spiky hemispheres, but they do not grow on the underside, so the short tail can be seen. The underside lacks spines along its entire length and may be covered with fine to coarse hairs or, in some echidnas of the hotter parts of Australia, may be almost naked.

Tiny muscle bundles are attached to the base of each spine that give the echidna control over the movement and direction of spines. Erect spines can anchor an echidna firmly in a log or other retreat and can even be used to facilitate climbing and uprighting if the animal has fallen or been placed on its back. When touched, an echidna will often hunch its shoulders and erect its spines, spiking the unwary inquisitor. Perhaps this behaviour is the source of the legend that echidnas 'throw their spines' (they don't).

Spines do not moult seasonally but can last many years before being lost. A common technique used by biologists in field studies or with captive echidnas is to mark individuals for identification with a piece of coloured tubing slipped over a spine. Even in the field, the spines marked with these tubes can remain in place for up to 10 years.

The skeleton

A mosaic of features

Many features of the monotreme skeleton are more like features of animals that lived in the Mesozoic than like those of living marsupial and placental mammals. In these features monotremes are similar to advanced mammal-like therapsid reptiles and early Jurassic mammals. It is remarkable that so many archaic features can still be discerned – some nearly unchanged – in the monotreme body plan that is perhaps 200 million years old. However, over that 200-million-year span monotremes have evolved many unique adaptations that overlie these archaic features. Most anatomists and palaeontologists agree that monotremes, therefore, show a mosaic of primitive and specialised features, as did many Mesozoic mammals.

Some features within this mosaic are specialisations related to the particular lifestyle of the echidna and the platypus. In some instances these specialised features are also seen in unrelated therian mammals with similar lifestyles. Such features are termed convergent. For example, echidnas have independently evolved skeletal adaptations that are similar to features in other ant-eating mammals such as the South American ant-bear group Xenarthra (see Figure 2.5).

It is difficult to sort out the primitive components of the monotreme skeletal mosaic because the main groups that might throw light on their evolutionary position simply are not well enough known to make comparisons. Late therapsids that might be direct ancestors and early mammals that appeared at about the same time as monotremes would be most useful but there is simply insufficient skeletal detail available in the literature. Therefore comparisons must be made with groups that are related to probable direct ancestors but are themselves mosaics with specialised features. Comparisons in this account are made in large part from descriptions of an advanced herbivorous cynodont (the tritylodontid, *Oligokyphus*); carnivorous African cynodonts; Late Triassic insectivorous morganucodontids *Eozostrodon* and *Erythrotherium*; and the docodont *Megazostrodon* (see Figure 2.2). The validity of such comparisons is supported by the fact that it was the skeleton of *Oligokyphus* that alerted researchers to the many shared, plesiomorphic features that therapsids and monotremes have in common. The fact that the jaw of *Teinolophos*, the oldest monotreme fossil, is only slightly more advanced than that of morganucodontids and docodonts, likewise supports the validity of comparisons between monotremes and these groups.

The skull and mandible

Monotreme skulls are basically mammalian in form. Both living monotreme families have ossified walls protecting the braincase, complete bony secondary palates and a single bone in the lower jaw, which are characters found only in mammals. Monotremes, however, have several plesiomorphic skull features not seen in most other living mammals, and these features have been used by many researchers as evidence for an ancient origin for the group.

The embryonic skull, which is cartilage before ossification, is of great interest in monotremes because of several plesiomorphic structures that appear during development of the embryo. Paired strips of bone (the *pilae antoticae*) arise from the base of the chondrocranium. Their function in monotremes is uncertain, and they appear to be vestigial. They are not present in any other living mammals, but they are found in reptiles and birds where they form a large part of the braincase. A cartilaginous scleral cup or ring supporting the eyeball develops in the chondrocranium and persists in the adult monotreme eye. This structure is likewise absent from other living mammals but occurs in reptiles and birds.

A small, sharp egg 'tooth' and an 'os caruncle', a bony knob that lies above the egg tooth, develop in embryonic monotremes (Figure 3.2). The so-called

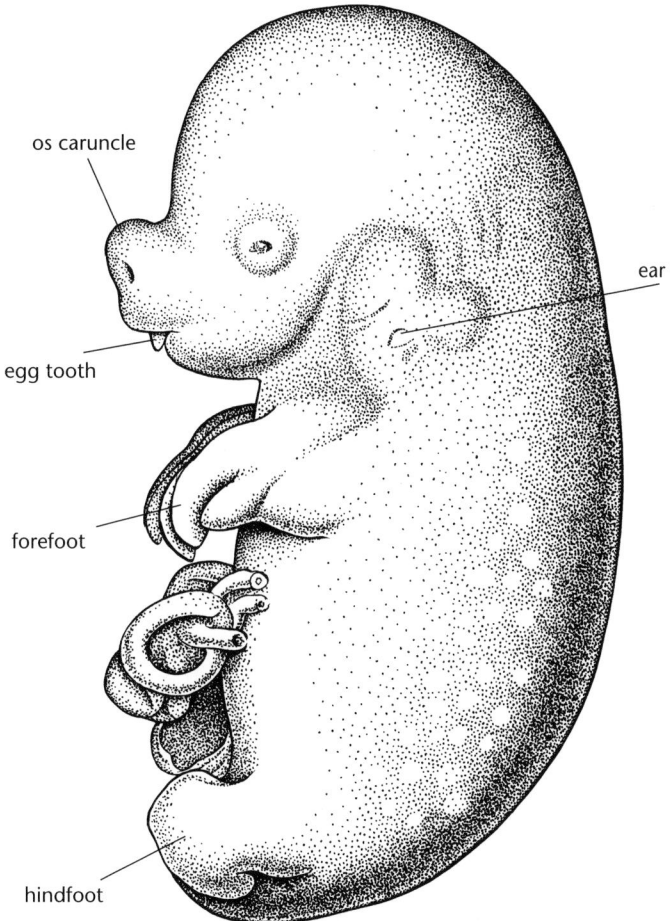

Figure 3.2 Echidna hatchling (after Semon 1897).

egg 'tooth' also occurs in birds and reptiles where it has the same function as it does in monotremes – to facilitate the emergence of the hatchling monotreme from the eggshell.

The adult echidna skull differs greatly in general form from that of the platypus. It is bird-like, with a large, domed cranium and long tubular snout (Figure 3.3). The snout is comprised of long, thin extensions of the premaxillary, septomaxillary and maxillary bones. The snouts in echidnas completely lack teeth. No indication of vestigial dentition has been detected in studies of echidna embryos. As far as the fossil record is concerned, echidnas have never possessed teeth, and no intermediate between a toothed monotreme and a prehistoric echidna has yet been found.

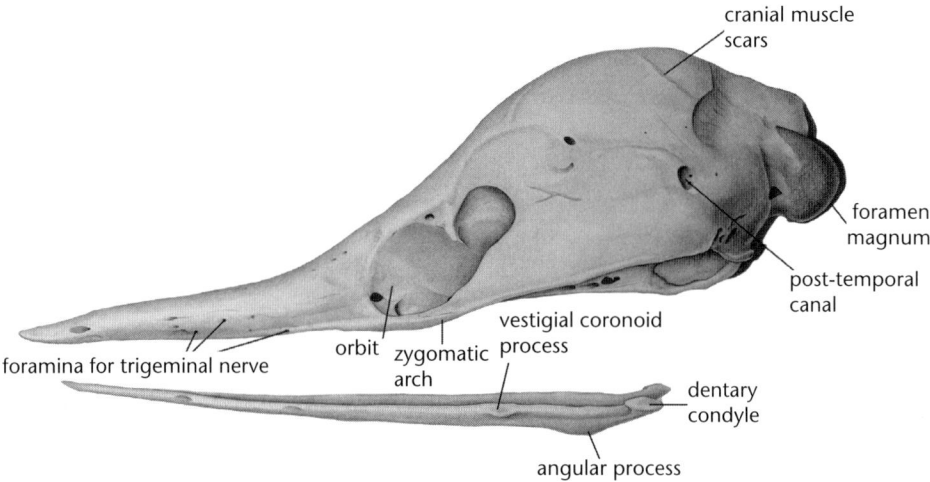

Figure 3.3 Later view of an echidna skull.

Several small holes for branches of the trigeminal nerve perforate the bony snout along its length. The nostrils are dorsally located and open at the anterior end of the snout.

The relatively enormous, hemispherical cranial cavity houses a surprisingly large brain (described in Chapter 4). On the floor of the cranial compartment there is a huge, horizontal cribriform plate perforated for the passage of numerous olfactory fibres leading from the olfactory epithelium in the snout to the olfactory bulbs of the brain. In both monotreme families the skull bones fuse together early in life, obliterating or obscuring suture lines. This makes the task of identifying the skull bones difficult and has led to some confusion in the identification of the component skull bones.

The dorsal and lateral sides of the skull roof bear scars for the origin of wide, strap-like muscles that run posteriorly from the skull to insert onto the shoulder, helping to give echidnas a somewhat neckless appearance. The zygomatic arches are thin and weak due to the reduction of the musculature required for chewing. The base of the skull is quite flat. The jugal, a small bone of the zygomatic arch of most mammals, is absent in echidnas although it is present in the platypus.

The tympanic region in monotremes, on the base of the skull, is quite primitive in configuration, which is to be expected given the primitive nature of ancestral monotreme jaws (see Chapter 2). The tympanic bone is not covered by a bulla (a bony acoustic and protective chamber) in either living monotreme family. The tympanic region in platypuses is completely exposed, but in echidnas a small overhanging lip of bone provides some protection for the middle ear. Monotremes have the three ear ossicles (see Figure 5.3, page 61) developed from reptilian jaw bones seen in therian mammals. Their structure is primitive, possibly because these bones were independently acquired in monotremes. The stapes is rod-like as in most marsupials and in the pangolin, *Manis javonica*. The tight articulation of the ear ossicles in echidnas contrasts with the lighter suspension in marsupials and placentals as well as in the platypus. This may be an adaptation for improving the conduction of sound through the body, especially the snout, to the Organ of Corti, where the cochlea terminates.

The cochlea is only partially coiled (see Figure 5.4, page 63), a three-quarter spiral rather than a full, or complete, 360° turn as in advanced therian mammals. The cochlear configuration has phylogenetic implications since a fully coiled cochlea is seen as a feature of advanced mammals. The recently excavated skeleton of an archaic therian mammal from China shows that it had a straight cochlea, suggesting that the partially coiled cochlea in monotremes may have developed convergently to the fully coiled cochlea of therian mammals.

As discussed in Chapter 2, archaic monotremes had a primitive mandible (lower jaw) with several accessory jawbones in addition to the dentary. Echidnas have lost all vestiges of this ancestral jaw form and, in fact, have lost almost all 'normal' landmarks on the lower jaw, making it probably the most reduced lower jaw of any mammal. The echidna mandible is comprised of two thin, rod-like dentaries, weakly fused at the symphysis and articulated loosely at the glenoid 'fossa', a shallow depression on the underside of the skull. The coronoid and angular processes are vestigial although development varies between individuals. One reason for such reduction is that the jaw in echidnas is not lowered, or opened, as in other mammals, including the platypus, but is instead rotated about its long axes, with the dentaries bound to the upper jaw by longitudinal ligaments running the length of the jaws.

The postcranial skeleton

There are seven cervical vertebrae in monotremes, as in other mammals, but they bear ribs, as they did in most therapsids and possibly in primitive mammals like *Megazostrodon*. In adult monotremes these cervical ribs fuse to the cervical vertebrae. The axis (the second cervical vertebra) of monotremes is similar to that of therapsids (*Oligokyphus*) in the shape of the body of the vertebra. The number of post-cervical vertebrae described in *T. aculeatus* varies. Cabrera in 1919 described 16 thoracic, 3 lumbar, 3 sacral and 12 caudal vertebrae; while in 1947 Gregory counted 15 thoracic, 3 lumbar and 2 sacral vertebrae. *Zaglossus* is described as having 17 thoracic, 4 lumbar, 3 sacral and 12 caudal vertebrae. There are not clear differences between thoracic-lumbar-sacral vertebrae in monotremes, a situation which may have led to the confusion in the literature over specific vertebral counts. The spinous processes between the first thoracic vertebra and the last presacral vertebra point posteriorly in *Tachyglossus*, as they do in *Oligokyphus*.

Monotremes have both vertebral ribs and sternal ribs joined together by strips of cartilage. The broad, overlapping sternal ribs are strongly ossified, as in birds, and attach directly to the sternum rather than by a cartilaginous connection, as is usual in mammals. The sternal ribs are a unique monotreme feature and are not found in this configuration in other tetrapods. Vertebral ribs in monotremes are not divided into a head and tubercle, as in other mammals, but are attached directly to the sides of the bodies of the vertebrae.

The pectoral (shoulder) girdle in monotremes is quite ancient and has no parallel amongst living mammals. The rigid pectoral girdle is composed of a series of extra bones that have been lost or reduced in nearly every mammalian

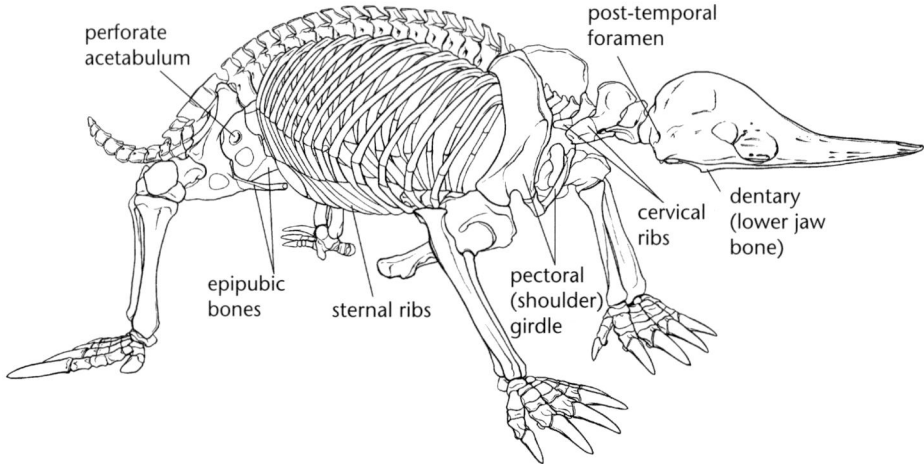

Figure 3.4 Skeleton of the echidna.

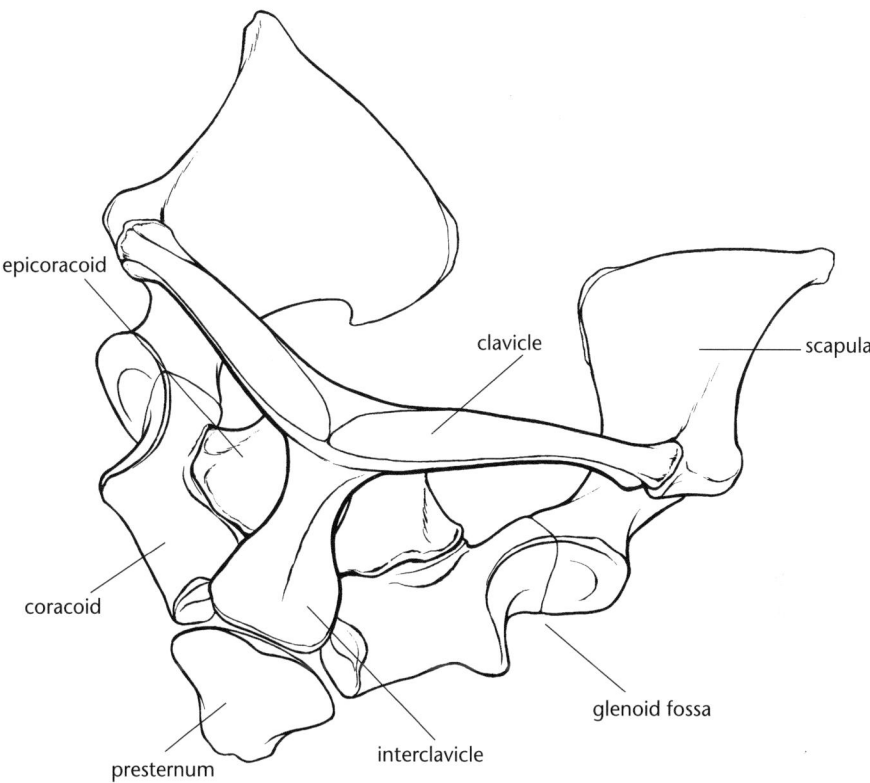

Figure 3.5 Echidna pectoral (shoulder) girdle.

group, living or extinct, for which the pectoral girdle is known. The configuration is similar to that of advanced therapsids like *Oligokyphus* and even more so to morganucodontids and docodonts, although monotremes have developed specialisations on this basic plan. The pectoral girdle consists of the following bones: scapulas at the back; coracoids, epicoracoids and clavicles at the sides; and a T-shaped interclavicle at the front (Figure 3.5). Monotremes are specialised in that the glenoid fossa on the scapula for reception of the head of the humerus is deep and laterally oriented for the horizontally oriented humerus. There is an articulation between the coracoids and interclavicle, providing additional rigidity. These are derived characteristics in monotremes and are probably related to digging habits. The monotreme scapula configuration on the other hand is primitive. The scapula is expanded dorsally and lacks an anterior recess for the attachment of the supraspinatus muscle, a feature well developed in therians but not in archaic mammals.

Modern mammals have lost all but the clavicle and scapula, resulting in a lighter suspension that permits greater mobility and freedom of movement in the shoulder. Fossil finds over the past decade show that the non-therian multituberculates and triconodonts, and archaic therian mammals (such as symmetrodonts), all had a more mobile and more advanced shoulder girdle than monotremes do. The retention of this rigid girdle in monotremes, particularly by the fossorial echidnas, probably reflects its utility in providing a strong and stable superstructure for the phenomenally strong forelimb musculature used in digging. The mechanics of this adaptation are discussed in Chapter 8.

In monotremes the short, extremely wide humerus is twisted on its axis and the distal end is expanded, as in some primitive mammals and therapsids but not as in therian mammals. Some features are very cynodont-like, including the shape of the humeral head, the relationships of the crests on the proximal end, and the form of the groove for the biceps muscle. The exceptionally wide distal end of the humerus serves as muscle attachment area for extensive forelimb musculature.

The humerus is held at right angles to the body and rotates about its axis during locomotion, rather than moving fore and aft, as in most other mammals where the limbs are held beneath the body. This arrangement may appear to be related to the sprawled stance of reptiles, but studies of echidnas have shown that this locomotory pattern is unique to monotremes. Both front and hindlimbs in *Zaglossus* are proportionately longer than they are in *Tachyglossus*; consequently *Zaglossus* holds itself higher off the ground than does *Tachyglossus*, perhaps to better navigate the complex topography of forest floors.

The manus (hand) has five digits with wide, spade-like claws, the hand thus forming a 'shovel' when the echidna digs.

The monotreme pelvic girdle is somewhat more advanced than the shoulder girdle although there are some very plesiomorphic features. The pelvis is short and broad and the iliac blade is directed anteriorly and dorsally, which is similar to that of the advanced therapsid *Oligokyphus*. Echidnas, intriguingly, have an unexplained perforation (incomplete symphysis) through the acetabulum as in birds, almost undoubtedly a specialisation. Epipubic bones (also known as marsupial bones) are attached to the anterior margin of the pubes, as in some advanced therapsids (tritylodontids), basal mammals, most marsupials and archaic placentals.

The short, stout femur in monotremes has several archaic features: the head is not inflected medially, the trochanters diverge laterally, and there is only a short neck supporting the head. This conformation resembles that of *Oligokyphus* and Late Triassic–Jurassic mammals that had not yet developed a stance in which the limbs were held beneath the body. In therian mammals and their ancestors, there are several evolutionary changes in the femur that redirect the leg under the body. The head develops a neck and is directed towards the midline of the body, and the trochanters migrate to accommodate the altered musculature. The monotreme configuration, like that of advanced therapsids and early mammals, has the femur (as with the humerus) projecting laterally (or close to it) from the body. Because of its horizontal orientation, the femur in monotremes rotates during locomotion. The femur in the platypus is more archaic in form than that of tachyglossids, and resembles *Oligokyphus* and that of morganucodontid femora in basic form.

The knee joint in monotremes has several derived features including a broad groove for the kneecap and flat, subequal, separated condyles on the femur. The flattened proximal end of the monotreme fibula is uniquely specialised in having an elongate proximal process. In tachyglossids the tibia and fibula are both rotated posteriorly so that the hindfoot is rotated outwards and backwards, an oddity that gives them quite a peculiar posture. Like the forefeet, the hindfeet are five-toed, but the claws are primarily used for grooming rather than digging. The 'big toe' is small; digits 2 to 5 are longer and more robust. The ankle joint is quite specialised, possibly derived from an ankle like that of Triassic mammals.

There are sharp, hollow, perforated spurs on the ankles in echidnas. These range from 0.5 to 1 cm in length. The spur is not connected in a functional way to the venom gland, unlike the crural system of the platypus where the venom, a neurotoxin, is delivered through the hollow spur to the unfortunate victim. The possession of ankle spurs has traditionally been considered a uniquely monotreme feature, but recent discoveries of near-complete Mesozoic mammal skeletons have shown that some archaic mammals also had such spurs.

4
The brain

For over a hundred years the echidna's brain has been a source of fascination and frustration for neuroanatomists. The total volume of an echidna's brain is a little smaller (about 25 ml) than that of a cat (30 ml). These two species are roughly equal in body size.

In 1902 G. Elliott Smith in his extensive work on the brains of mammals stated:

> 'The most obtrusive feature of this brain [echidna] is the relatively enormous development of the cerebral hemispheres, which are much larger, both actually and relatively, than those of the platypus ... The meaning of this large neopallium is quite incomprehensible. The factors which the study of other mammalian brains has shown to be the determinants of the extent of the cortex, fail completely to explain how it is that a small animal of the lowliest status in the mammalian series comes to possess this large cortical apparatus. In other small, terrestrial, insect-eating mammals such as the Pangolins and the Anteaters, and in the fossorial Bandicoots, Hedgehogs and Armadillos, we find highly macrosmatic brains [having an acutely developed sense of smell] with small neopallia: and yet in Tachyglossus,

whose mode of life is not dissimilar to many of these mammals, we find alongside the large olfactory bulb and great pyriform lobe of the highly macrosmatic brain a huge complicated neopallium.'

Smith's claims about the absolute and relative size of the echidna brain compared with that of the platypus were not supported by quantitative data. More recently specific measurements, albeit on only one specimen of each species, revealed that although the echidna's brain volume was 3.1 times greater than that of the platypus, the platypus had a larger brain volume to body mass

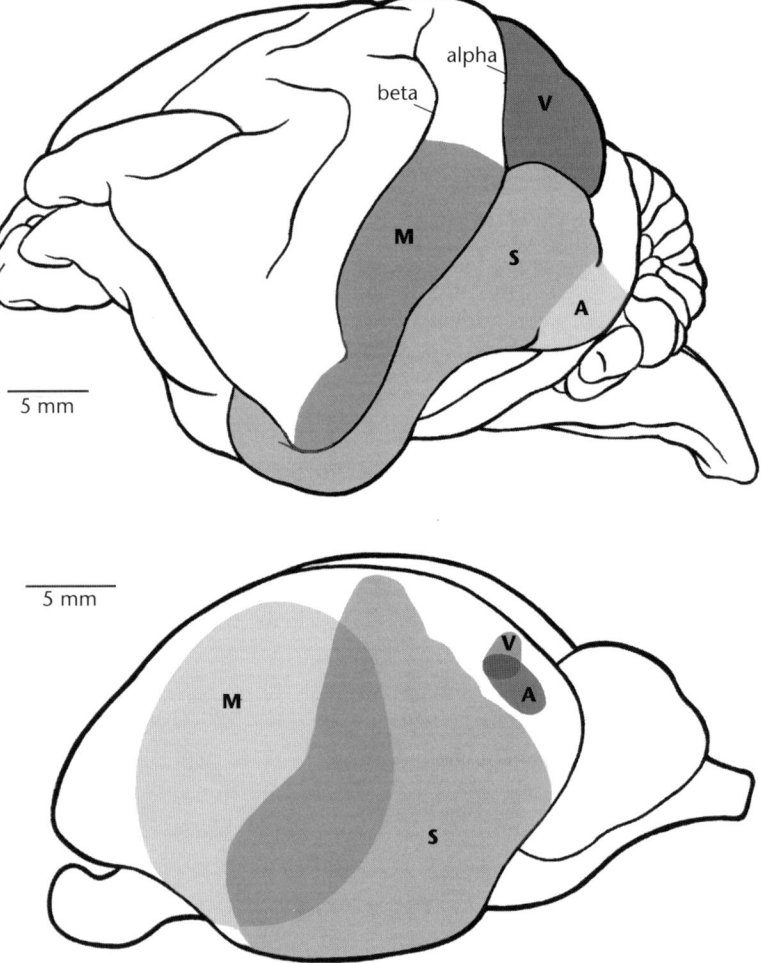

Figure 4.1 Left side view of an echidna brain (top) and platypus brain (bottom) showing the cortical motor and sensory areas and the alpha and beta sulci as defined by Smith 1902. A = auditory area, M = motor area, S = sensory area, and V = visual area.

ratio. The neocortex, that is, the non-olfactory portion of the cerebral cortex, represented about the same percentage of total brain volume in the two species, 48 per cent in platypus and 43 per cent in echidna. Clearly more data are needed to unequivocally substantiate these observations.

If we simply look at the surface of the brain of these two monotremes, we notice at once that the platypus brain is largely smooth (lissencephalic) whilst the echidna's brain is elaborately folded with marked fissures between these folds (gyrencephalic) (Figure 4.1). These convolutions enable a greater area of cortex to be accommodated in the skull. Indeed the percentage of both area and volume of the neocortex buried in fissures is approximately as large in the echidna (36 per cent) as it is in a domestic cat (40 per cent) and a squirrel monkey (39 per cent). Such a convoluted cortex is generally considered to indicate a more neurologically sophisticated mammal such as a primate or a carnivore.

Although the cerebral cortex is thinner in the echidna compared with the platypus, the brain cells are generally bigger and twice as densely packed and may reflect a higher grade of cerebral organisation. The greater mass of nerve fibres carrying impulses to and from the cortical cells must account largely for the greater bulk of the cerebral hemispheres. This marked difference in cerebral anatomy suggests that the evolutionary pathways of echidna and platypus must have been separated for a long time.

In the echidna, areas of the brain directly related to body movement, body sensation, hearing and vision are restricted to the back halves of the cerebral hemispheres whereas in the platypus these areas extend over a large part of the surface of the brain including the frontal cortex. The positions of the sensory, visual and auditory areas in the echidna are unlike those described in any other mammal. The relationships of these areas has been described as 'a rotational dislocation of the areal relations found in placentals …'. Such comparisons are often made between echidna and platypus but the research involved is complex.

Almost half of the echidna's sensory area is allocated to the snout and the tongue (Figure 4.2), which is not surprising in an animal that depends so much on these structures for obtaining food. The trigeminal nerve is greatly enlarged as are the related nuclei in the brain stem which is consistent with the presumably huge sensory input from the snout (see Chapter 5). The olfactory bulb in the echidna represents around 3.1 per cent of the brain volume compared with 0.8 per cent in the platypus. The paleocortex, that is, the phylogenetically older portion of the cerebral cortex, is also relatively greater. Both these structures are related to the sense of smell. Conversely the cerebellum is relatively greater in the platypus. These findings may reflect the fact that smell is more important to the terrestrial echidna but relatively unimportant to the aquatic platypus. A greater capacity for processing sensory information from the vestibular

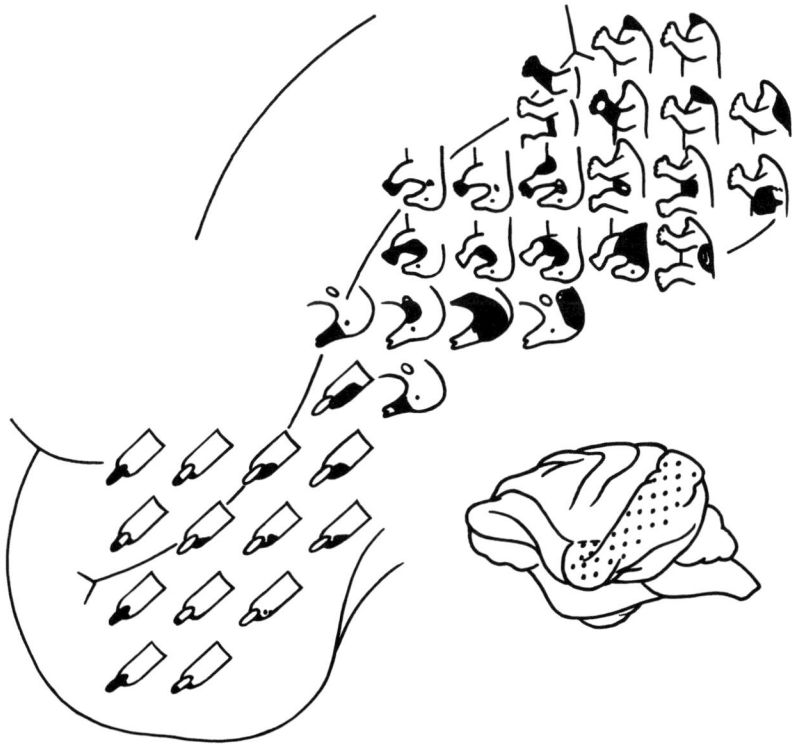

Figure 4.2 Cortical sensory representation in the echidna's brain. Note the large area taken up by the snout and tongue. Only the sensory area (shown as stippled on the smaller diagram) is enlarged above.

organs may be more important to the platypus swimming in the water than an echidna walking on land.

Perhaps the strangest feature of the echidna's brain is the extent of the prefrontal cortex. In humans, stimulation of the front 29 per cent of the cerebral cortex does not produce any observable motor response or sensory perception and has therefore been called 'the silent area'. In the echidna this area takes up a remarkable 50 per cent of the cerebral cortex, a greater proportion than any other animal. What then is the function of the prefrontal cortex in the echidna?

In humans the prefrontal cortex is believed to be involved in future planning, construction of alternative interpretations of an event, detection of novelty and monitoring behaviour. To perform such a role would naturally require the association of incoming sensory information. Sensory information is projected via the thalamus to the prefrontal cortex. In general, the thalamocortical relationships in echidna are similar to those of placental mammals. One unusual feature is the relatively large mass in the dorso-fronto-medial thalamus which projects to the frontal cortex. It is believed that this large nuclear region is, in part, the equivalent of the mammalian mediodorsal nucleus and may be the thalamic component that processes olfactory information destined for the cerebral cortex. The processing of information from electroreceptors has also been suggested as a function of the frontal cortex. It has been questioned whether this area of the echidna's brain can really be considered as homologous with the same area in human beings and it has been suggested that this large expansion of the frontal lobe is part of the olfactory specialisation.

Consideration of behavioural aspects of the echidna's world may give clues to elucidating this puzzling situation without the necessity to write off the echidna's prefrontal cortex as an embarrassing inconsistency. The extensive work by Max Abensperg-Traun and colleagues in 1991 on the relationship between the echidna and its prey species shows that echidnas generally adjust their foraging effort in response to prey abundance. It has been suggested that prey defences deter echidnas which results in the recovery of prey colonies after raids, thus ensuring a permanent supply of food. However, at times echidnas seem undeterred by prey defences and can override external 'controls' on food conservation. Is it possible that echidnas have the capacity to exploit their prey in a planned fashion? Such 'farming of prey' would require a detailed cerebral map of the variety of prey species, their location, depth, density and nutritional value. The influence of the time of the year and air temperature on prey behaviour would also be important factors to be taken into consideration. Sophisticated processing in the frontal cortex of this wealth of data could be essential for forward planning, in particular detection of changes in the external environment and their correlation with internally stored models of the same. These functions in the echidna might represent what we would call conscious awareness, including rumination on the past and speculation about the future; that is, an awareness of the passage of time or consciousness of a temporal stream.

Recently psychologists have begun to question the idea that animals do not possess the capacity for conscious thought. Previously it had been implicitly believed that animals live their entire lives as 'sleepwalkers'. Animal thoughts may well be limited to simple matters such as food, predators and social companions. However, thinking about alternative actions is much safer than trying them first in the real world where mistakes could be fatal. As we have already noted above, the echidna possesses a massive frontal cortex where we believe thought takes place. Echidnas spend much time resting in dark hollows, logs and burrows. If echidnas think, what do they think about? Perhaps they have invented a variety of thought games such as cerebral chess, where the opposing forces are ants and termites with delectable queens ruling soldier knights and indigestible pawns.

Learning and intelligence

How intelligent is the echidna? To examine this question we must devise methods for measuring relevant intelligent behaviour. We can look at the echidna's behaviour in both natural and laboratory environments. An echidna is quite capable of coping in a wide variety of environments. It is extremely efficient at finding its prey even though this might be deep in logs or underground (see Chapter 5). It knows the boundaries of its home range and usually keeps within these limits even though this may be a complex environment. However, little is known about the echidna's ability to learn in its natural habitat. Observations have been made of behaviour of animals in zoos but this situation can be so artificial that conclusions may be quite misleading.

Psychologists have developed techniques for investigating aspects of intelligence in the laboratory. A simple device which can be used to determine an animal's ability to learn is a 'T' maze (Figure 4.3). A food reward is placed in one arm of the 'T' and the animal has to learn over a number of tests which arm contains the food. Once the animal has learned this task the food can be switched to the opposite arm of the 'T' and the number of trials taken to learn this new position is determined. Echidnas are quite proficient at these tasks and demonstrate a learning capacity similar to other mammals.

In another experiment the echidnas had to press one of two pedals with the forefoot to receive a food reward. When first introduced to the experimental

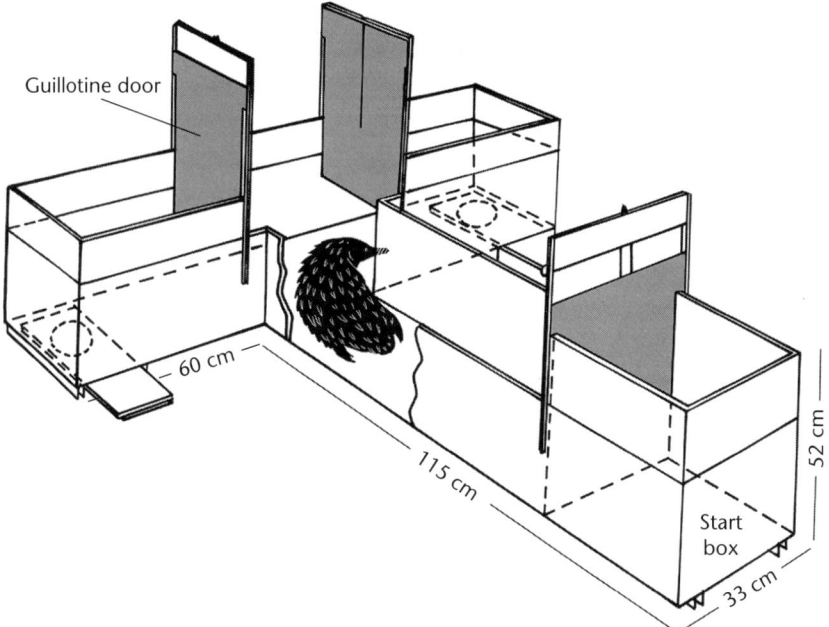

Figure 4.3 T-maze apparatus used in intelligence testing of echidnas.

room the animals investigated their environment by sweeping the snout from side to side, rearing up the head and walking about. All but one of the six animals quickly discovered, without specific training, that pressing the pedal resulted in the supply of food. The pedal which gave a food reward could be alternated and in some experiments the pedals could be distinguished by their surface being either black rough sandpaper or a white card under smooth Perspex. Visual and tactile cues were combined because at the time echidnas were considered to have poor vision. Using combinations of these variables, a variety of tasks could be set up. Unsuccessful responses were often associated with vigorous kicking of the pedals. The results showed that echidnas are able to store, classify and integrate visual and touch information, indicating a degree of attention characteristic of a highly organised nervous system. This process is believed to be carried out in areas of the brain like the frontal cortex. Retesting of three echidnas indicated that their memory extended at least one month. An in-depth study on echidna sight revealed that echidnas are able to use their vision alone in order to learn how to obtain food rewards.

The outcome of these studies indicates that the learning ability of an echidna is similar to a rat or cat. Of course the measure of intelligence determined from these studies could be biased by the type of test employed. For example, an echidna might exhibit a degree of intelligence completely beyond our capacity if it used its snout to detect small electrical fields. Buchmann and Rhodes summarised our present state of knowledge succinctly when they wrote:

> 'Further studies of learning will undoubtedly disclose important facts about the intelligence of these remarkable animals and modify the quaint, explicitly and tacitly-held views that echidnas are little more than animated pin-cushions or, at the best, glorified reptiles.'

Echidna sleep
The function of sleep and dreaming is still largely unknown. One possible path to a better understanding of these phenomena may be to ask the question: 'How did the biological system we call sleep evolve?' Since monotremes are believed to have diverged from the placental and marsupial lines very early in mammalian evolution, it might be expected to provide some clues about the evolution of sleep.

Sleep in mammals can be divided into two stages called rapid eye movement (REM) and non-rapid eye movement sleep (non-REM). REM takes its name from the rapid eye movements that occur during a particular state of sleep. In humans dreams are frequently reported in REM sleep. REM is also called paradoxical and dream sleep. The cortical electroencephalogram (EEG) during REM is of low voltage with rapid bursts of activity. During this sleep state tone in most skeletal muscles is abolished. Heart rate and respiration become irregular and body temperature tends to drift towards ambient, as though homeostatic control is reduced. Brain metabolic activity is high and its temperature rises during REM in most mammals. REM sleep is generated in the brain stem by bursts of firing from the reticular formation. This discharge produces the twitching that characterises REM sleep.

In contrast, in non-REM or slow wave sleep the voltage of the cortical EEG increases. The cortical neurons fire rhythmically and in synchrony with neighbouring neurons. Heart rate and respiration become less variable. Metabolism is greatly reduced in the cortex and throughout the brain during non-REM sleep. Brainstem neuronal activity is also markedly reduced.

Nearly 30 years ago a careful and detailed study by T. Allison and his co-workers profoundly influenced our concept of sleep in the echidna. This investigation found that the echidnas spent about 93 per cent of their sleep time in slow wave sleep which resembled the non-REM sleep state seen in placental and non-marsupial mammals. In the remaining sleep time the electroencephalogram briefly showed patterns that looked similar to paradoxical sleep but the experimenters were unsure about their true nature. Consequently they referred to these periods as 'PS?'. They could not define REM sleep because the echidnas had the annoying habit of burying themselves in corn-cobs that had been supplied to provide a stress-reducing environment. Consequently the eyes could not be observed during sleep. In addition, when the animals became quiet before sleep, neck muscle tone disappeared and so it was impossible to detect any further reduction in tone which might have indicated REM sleep in the echidna. None of the other measured variables like heart rate or respiration provided changes that were completely supportive of paradoxical sleep.

Since the electrical origin of REM sleep appears to be in the brainstem, more recent work has examined the electrical activity in this region of the echidna brain. The pattern of activity of brain stem neurons during sleep in the echidna resembles that seen in REM sleep of placental mammals.

REM sleep has been described in the platypus in combination with eye, head and neck twitching. In 1860, the zoologist George Bennett reported that very young platypuses show swimming movements of their forepaws while asleep, indicative of dreaming. No detailed study of sleep in the platypus has been reported to date.

The most recent work on echidna sleep appears to have uncovered the reason for the past confusion over whether the echidna has REM/paradoxical sleep or not. REM sleep is suppressed by stress, particularly heat and cold. Sleep studies were therefore performed on echidnas at thermoneutral (25°C), low (20° and 15°C) and high (28°C) ambient temperatures. By using stringent sleep-scoring criteria and refined data analysis, these experiments demonstrated clearly that REM sleep does occur in echidnas at an ambient temperature of 25°C, but is partially or totally suppressed by temperatures several degrees above and below this level. Previous studies did not control for ambient temperature. Importantly, the environmental temperature in Allison's experiments is estimated to have been 21°C which would have almost completely eliminated REM sleep.

The characteristic features of REM sleep have now been demonstrated in the echidna and are similar to those in all previously investigated mammalian and avian species, excluding the dolphin. Thus REM sleep either evolved twice independently, that is, in birds and mammals, or, more probably, only once in a common reptilian ancestor. The origin of REM sleep in the brain stem, which is a phylogenetically ancient part of the central nervous system, and the apparent inhibition of thermoregulation during REM sleep, supports the concept that REM sleep is an ancient form of sleep. Fundamental relationships between the control of sleep, temperature and hibernation may yet be revealed in the echidna.

Spinal cord

The echidna's spinal cord is of relatively shorter length than that of a human. In the echidna it terminates at the level of the 7th thoracic vertebra whereas in the human it ends between lumbar vertebrae 1 and 2. It has been suggested that this shorter length in the echidna allows it to roll up into its hyperflexed defensive posture without over-stretching the spinal cord. A greater proportion of the stretching is presumably taken up by the possibly more resilient sheaf of nerve roots of the *cauda equina* which radiate from the distal end of the spinal cord.

The arrangement of the nerve cells in the echidna's spinal cord is similar to eutherian mammals. It is interesting to note however that the motor or corticospinal nerve tract crosses from one side of the brain stem to the other in the pons. In no other mammal does this so-called pyramidal decussation occur as high as in the echidna. Since comparative data is not available for the platypus, it is not known if this is a monotreme character or an echidna specialisation. The motor nerve tract crossover usually occurs in the medulla. Intriguingly, a high decussation is also found in a small number of highly specialised mammals including the pangolin and the armadillo. The relevance, if any, of these observations has yet to be elucidated. The corticospinal tract in the echidna extends down the greater length of the spinal cord which is a feature of advanced neurological organisation as seen in primates and carnivores, and allows direct cortical control over motor neurons down to the tail and hindlimbs. Hedgehogs and tree shrews have a termination of the corticospinal tract at a considerably higher level than that seen in the echidna.

5
Senses

Vision

Situated near the base of the snout, the echidna's eyes appear black, small and somewhat protruding. The eyeball is roughly spherical and about 9 mm in diameter. The visual system of the echidna is an unusual mixture of both reptilian and mammalian characteristics. For example, the cartilaginous layer just below the fibrous scleral layer of the eyeball is characteristic of birds and reptiles. Six extrinsic eye muscles move the eye in its socket. The arrangement of the superior oblique muscle in monotremes was described by Walls in 1942 as wholly mammalian:

> 'In echidnas there is a slip which runs from the old sub-mammalian origin (on the anterior nasal orbital wall) to an insertion on the globe, but merging in this same insertion is a second slip, muscular almost to the globe, which comes through a pulley from an origin only a few millimetres anterior to the deep point-of-origin of the four recti. The duck-bill has only this long portion, and moreover has it as in higher mammals, i.e. originating with the recti and becoming tendinous before reaching the pulley, with the latter chondroid rather than soft as in the echidnas.'

Walls intimates that the evolution of the elongated superior oblique muscle may have been associated with the need for binocular vision. The oblique muscles are required for the correct viewing of an object when the head is moved laterally in order that the image falls on the same part of the retina in both eyes.

Cornea

The small flat cornea occupies about a sixth of the circumference of the eyeball (Figure 5.1). The surface layers of the cornea are keratinised, that is, the constituent cells lose their nuclei and are converted into a horny material known as keratin. This kind of change is unusual in human eyes but occurs in aardvarks and some armadillos that eat ants. Such hardening of the surface of the cornea presumably protects the sensitive underlying cornea from irritation and injury due to ant bites, defensive chemicals secreted by ants and termites and debris during digging. In addition, it might give some added resistance to penetrating injury by its own or another echidna's spine. Such impalement has been reported to have occurred when an echidna defended itself by rolling up into a ball in characteristic fashion and punctured its left eye with its own tail spine. Such eye injury might also occur during aggressive encounters between echidnas.

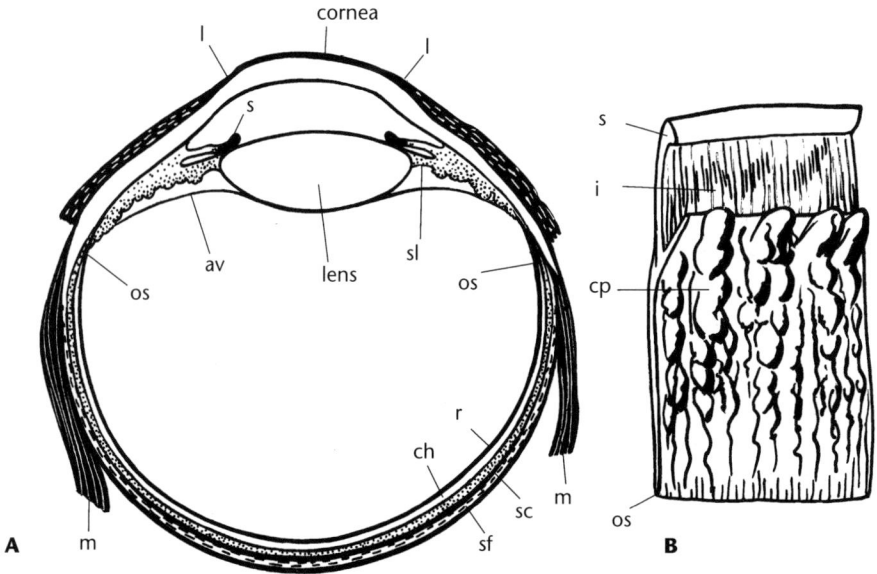

Figure 5.1 (A) Transverse section through the eye. (B) Segment of the posterior surface of the iris and ciliary body. Abbreviations: av = anterior surface of the vitreous humour, ch = choroid, cp = ciliary process, i = iris, l = lateral limit of the cornea, m = rectus muscle, os = ora serrata retinae (the anterior limit of the retina), r = retina, s = sphincter, sc = scleral cartilage (black), sf = fibrous layer of sclera, sl = suspensory ligament.

The iris consists of little more than two heavily pigmented layers. There is no dilator muscle but a massive constrictor muscle lies at the edge of the iris around which the pigmented layers are rolled. This iridal pigmentation gives the echidna eye its characteristic black 'beady eyed' appearance.

Ocular accommodation

The echidna has the flattest of all lenses with a flatness index (diameter divided by thickness) of 2.75. This value is only approached by some higher primates including humans, with a flatness index of approximately 2.7. Such a lens shape is required for the clear viewing of distant objects. Rays of light from a distant object are effectively parallel and the lens is set to focus these rays on the retina. For close objects the focal point of such a flat lens would fall behind the retina and consequently these objects would not be in focus. To deal with this problem marsupials and placentals possess a ciliary muscle in the eye. Contraction of this muscle releases the tension in the suspensory ligament of the lens which allows the lens to become more curved due to its inherent elasticity. Consequently the eye accommodates to allow clear vision of close objects. However the echidna does not appear to possess a ciliary muscle.

Some interesting observations on accommodation have been made by Richard Gates. In order to examine the echidna retina, Gates found that he had to anaesthetise the animals with halothane. Conscious echidnas tried to avoid the light of the ophthalmoscope, squeezed the eyelids shut and struggled vigorously. Early in the recovery from anaesthesia the echidnas' eyes protruded markedly, becoming less so as the anaesthetic wore off. Simultaneously the accommodative capacity of the eye decreased during recovery. Without a ciliary muscle to change the shape of the lens, Gates argued, perhaps the extrinsic eye muscles can act in such a way as to elongate the echidna's eyeball and in this fashion produce accommodation when the echidna is viewing close objects. He likened this mechanism to the old method used to alter the focus of a camera in which the distance between the photographic plate and the lens was varied. Examination of a normal awake echidna in its natural state revealed that, while eating or approaching close-up objects, the eyes protrude forward in a manner similar to that found in his animals awakening from anaesthesia.

Retina

Viewed through an ophthalmoscope, the retina is a uniform lavender colour and no blood vessels or fovea are apparent. The monochromatic appearance of the retina is similar to that found in fish, amphibians and the hairy armadillo. Oxygen and nutrients are supplied to the retinal cells from the substantial choroidal vessels. This complete nutritional dependence of the retina on the choroid is characteristic of light-shunning vertebrates.

Echidnas are known to forage both day and night. There has been considerable debate about whether the visual receptors in the retina are purely of the rod type which provides black and white vision and is particularly effective in dim light (scotopic vision). It is clear now that there are also a smaller number of cone receptors which are involved in vision in bright light (photopic vision) and the perception of colour. In the echidna eye, cones constitute 10 to 15 per cent of the photoreceptors, whereas in the human eye only about five per cent of the photoreceptors are cones. However, the distributions of the photoreceptors over the echidna retina show little regional specialisation. There is considerably less centralisation of cones in the retina of the echidna compared with most other mammals. The 2:1 centre-to-periphery ratio of the cones in the echidna retina is very much less than that seen in the monkey (100:1) and the cat (7.5:1). Nothing is known about the echidna's capacity for colour vision. It remains to be determined if subtypes of cones can be identified in the echidna, based on their visual pigment content.

Optic nerves and axis

It has been estimated that there are approximately 15 000 nerve fibres in each optic nerve of the echidna. The vast majority of the echidna's optical nerve fibres cross over to the other side of the brain at the optic chiasma. Only about one per cent (that is around 150) fail to do so.

In humans the eyeballs are set in the skull in such a way that they look straight ahead, that is, the optical axes are parallel and in the resting state do not diverge from the median line of the skull. The fibres from the nasal halves of the retina cross to the opposite side of the brain, whereas the fibres from the temporal (lateral) halves do not cross over. Thus approximately 50 per cent of fibres cross over. A viewed object forms a retinal image in both eyes and fusion of these images allows binocular or three dimensional vision. However it has been stated that in many mammals, with laterally directed eyes and therefore limited binocular vision, the degree of decussation is much greater, so that in the rat, for example, practically all of the optic fibres pass to the opposite side of the brain.

There appears to be some confusion in regard to the optical axes of echidna eyes, but most writers report the eyes to be directed forward to some degree. A hundred years ago George Lindsay Johnson designed a goniometer which allowed him to measure the divergence of the optical axes from the median line of the skull in a wide range of mammals. Most of his work was performed in the Gardens of the Zoological Society of London and 'Through the kindness of some friends in Australia' he obtained 'several living echidnas'. Using his goniometer he found that 'the higher the order [of species], the nearer the axis approaches parallel vision …'. Johnson was surprised to find that the echidna ranked so high in the list of divergences with an angle of only 25 degrees from the median line. This finding indicates that there is significant overlap of the visual fields of the two eyes and a degree of binocular vision. Despite the very small number of uncrossed fibres and lack of a corpus callosum, several other possible pathways exist in the echidna for the exchange of visual information between the cerebral hemispheres.

Early researchers suggested that the echidna has poor visual ability. However, Gates was able show that echidnas could learn simple visual discrimination between black and white, as well as between vertical and horizontal lines. His experiments suggest that echidnas have a visual capability which is probably equivalent to that of the brown rat (*Rattus norvegicus*). Echidnas clearly have a visual capacity somewhat better than 'dismal' as has been occasionally claimed.

However, certain facts need to be taken into consideration. First, much of its prey is usually not immediately visible to the echidna. Ant nests are below ground, and termites reside within earthen mounds, logs and trees. Second, echidnas are quite capable of foraging in the darkness of night. Third, blind echidnas remain healthy despite their complete lack of vision. Both partially and completely blind echidnas are known to have remained healthy in their natural environment for extended periods of time. Clearly one or more sensory inputs other than vision allow the echidna to detect and obtain its prey.

Hearing

Pinna and external auditory canal

Each pinna consists of a single undifferentiated plate of cartilage embedded in the adjacent skin and muscle (Figure 5.2). The ear is ovoid in shape with its long axis running dorso-ventrally. At the ventral end of the pinna the lateral margins of the pinneal cartilage merge forming a goblet-like structure. Consequently the pinneal opening is roughly triangular in shape with the base dorsally and the apex near the ventral end of the ear. Hairs also grow from the internal surface of the pinna.

The external auditory canal consists of cartilaginous rings joined by a membrane in a fashion similar to the trachea. The canal runs a tortuous path. Its first part takes off at right angles from the pinneal cavity, runs anteriorly, is elongated and contains hairs. The shorter second and third parts run medially and dorsally respectively, finally reaching the opening in the skull. The canal is widest at the pinneal end and narrowest at the cranial end. It is likely that muscles associated with the outer ear serve to close it off by pressing the pinna flat and approximating the inner walls of the external auditory canal. Photographic evidence indicates that during normal ambulation the outer ear is closed but upon approaching a strange object the spines adjacent to the outer ear are raised thereby opening the external auditory canal.

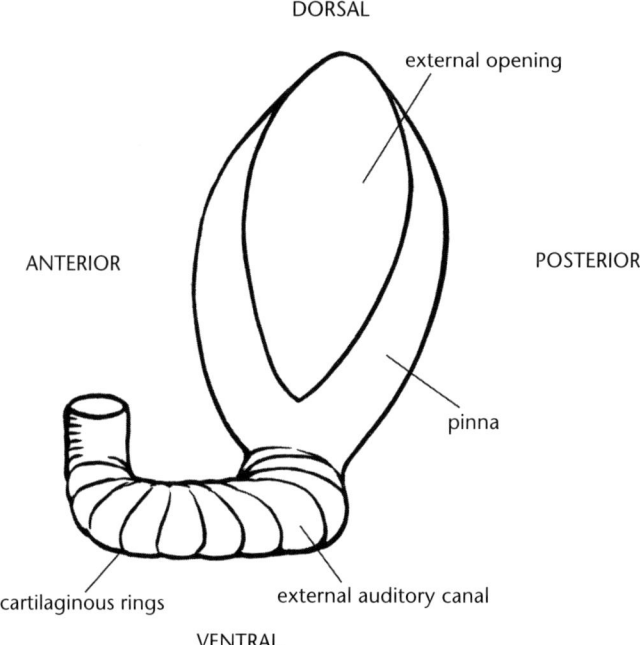

Figure 5.2 Exposed framework of the left pinna and external auditory canal viewed from the side.

The fact that the echidna pinna does not protrude means that a predator cannot seize it during an attack. Such an arrangement also reduces the possibility of trauma to the ear during burrowing. Hairs, both surrounding the pinneal opening and within the pinna, greatly reduce the risk of invasion of the external auditory canal by debris and prey species such as ants and termites. On the other hand, the ear provides a relatively protected environment for ticks.

The intramuscular nature of the pinna and its relative absence of morphological differentiation reduces its ability to direct airborne sounds into the acoustic meatus. The ear canal goes through three right angle turns which include all three spatial planes. This angularity and narrowness of the canal may protect against high intensity sounds. The canal is an entirely extracranial structure and as such the transmission of bone-conducted sound to the canal would be negligible. Consequently, airborne sounds would be impeded and the purity of bone-conducted sound to the inner ear enhanced. The significance of these adaptations becomes clearer upon examination of the middle and inner ear.

The middle ear

The external auditory canal ends at the tympanic membrane which separates the outer ear from the skull-encased middle ear cavity. Sound is progressively conducted through the middle ear by a chain of three small bones, in the following order: malleus, incus and stapes. This is characteristic of mammals. The footplate of the stapes stands on the membrane of the oval window of the cochlea. The oval window in the echidna is not oval but round.

The malleus bone is large and attached to the tympanic membrane and rigidly fused to the next bone in the chain, the incus. The malleus and incus are tightly attached at their most lateral corner to the periotic bone of the skull (Figure 5.3). Application of a small rod to the malleus demonstrates that quite considerable pressure produces little movement of the middle ear bones. This stiffness of the ossicular chain appears to be due to the tightness of the attachment of the incus and malleus to the skull. The velocity and amplitude of movement of the stapes when sounds over a range of frequencies from 100 Hz to 24 KHz (with intensity kept constant at 100 dB) were introduced to the external auditory meatus have been measured. In the echidna ear both measurements are

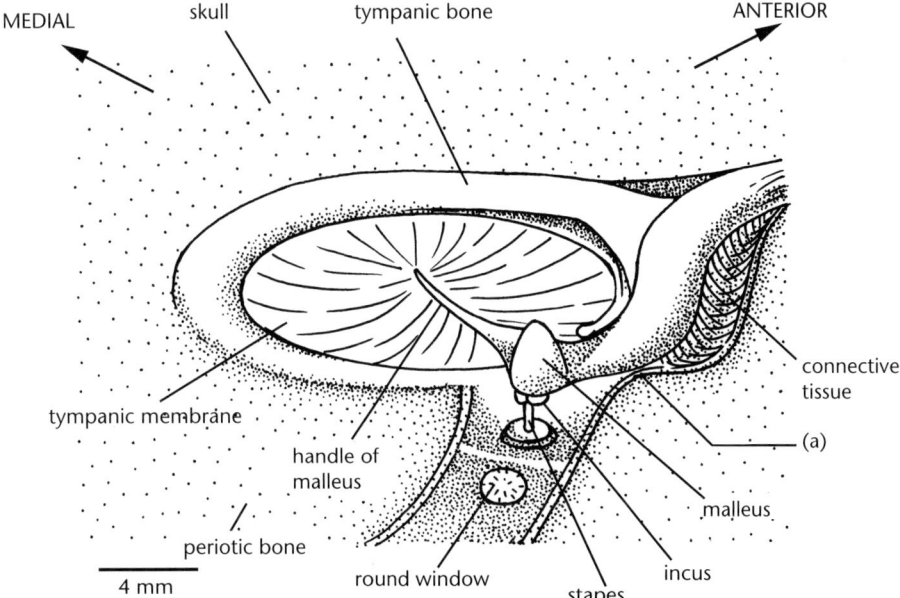

Figure 5.3 Structures of the left middle ear when viewed from a dorso-lateral position. (a) indicates the point of tight attachment of the incus and malleus to the skull.

substantially less at low frequencies than comparable data obtained from a placental mammal, the guinea pig, and a reptile, the dragon lizard. Only when the frequency is increased to 6 KHz is the stapedial function of the echidna similar to the other animals. The reason for the poorer function at low frequency presumably lies in the considerable stiffness of the middle ear system. So tight is the connection of the malleus to the adjacent skull bone that it seems to function as though it were part of the periotic bone acting as a supporting structure for the tympanic membrane rather than as an auditory ossicle. It is quite unlike the lightly suspended mallei of some placental mammals. The system appears to be most inefficient in conduction of airborne sounds and more likely suitable for the transmission of bone-conducted sounds to the inner ear. Supportive evidence for this hypothesis is the observation that a tap on the snout of the echidna elicited a large potential at the round window.

Modern non-invasive techniques used to measure hearing sensitivity in lightly anaesthetised wild-caught echidnas have revealed that the echidna has a significantly narrower frequency range, but within that range the sensitivity is comparable to that of typical therian mammals such as the rabbit and gerbil.

Inner ear: the cochlea

The cochlea of the inner ear in most mammals is coiled like a snail shell and in humans has two-and-three-quarter turns. In the echidna the cochlea has a banana-shaped curve with only a three-quarter spiral turn (Figure 5.4). Despite the external appearance, which is reminiscent of avian and reptilian cochlae, the internal structure of the echidna's cochlea is essentially mammalian with the usual three scalae or passages. The organ of Corti is basically mammalian and is approximately 7.6 mm in length, much shorter than in eutherian mammals. Movement of the oval window results in fluid movement in the scala vestibuli, distortion of the basilar membrane and movement of the tectorial membrane against the hair cells. The hair cells in turn convert this mechanical movement into nerve impulses. Detailed examination of the structure of the organ of Corti by electron microscope reveals that the echidna has almost as many inner hair cells (2700) as in humans (3000), although the echidna has a larger number of rows of cells with a smaller density of cells per row. The inner hair cells in the echidna appear to give rise to most of the sensory output from the organ of Corti. The number of outer hair cells is considerably less in echidnas (5050 compared with 12 000 in humans), but comparable to the subterranean mole rat. Outer hair cells may play an active role in facilitating extended high frequency range in other eutherian mammals which may not be so relevant for the needs of the echidna.

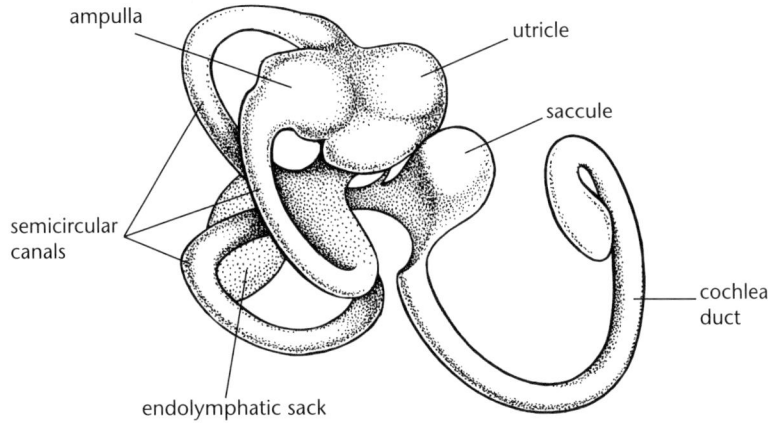

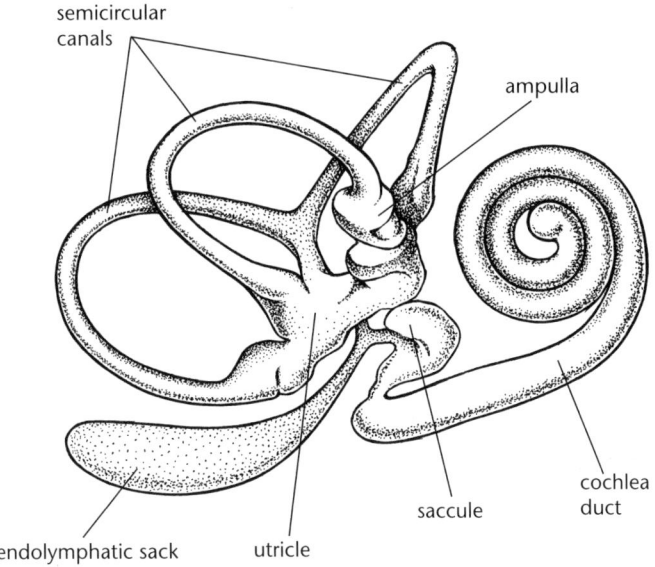

Figure 5.4 Structures of the right inner ear of echidna (top) and placental mammal (bottom) viewed laterally. Auditory receptors are located in the cochlea duct, while vestibular receptors are located in the ampulla, utricle and saccule.

The echidna cochlea responds to a wide range of frequencies with the most sensitive frequency lying close to 5 KHz, but its sensitivity is poor compared with most eutherian cochleae. It has been suggested that a cochlea most sensitive to frequencies near 5 KHz is just the thing for detecting noises emitted by termites or for detection of vibrations transmitted through bone. Such vibrations could reach the middle ear via the snout and elongated jaw bone. Measurement of the sound spectra produced in the ground by ants and in timber by termites could be valuable in further illuminating this matter.

Inner ear: the vestibular organs

The other major component of the inner ear is the vestibular system which consists of a number of sensory organs contained in the three semicircular canals, in two dilatations called the utricle and the saccule, and in the tip of the cochlea, the lagenar macula (Figure 5.5). The semicircular canals are mounted at right angles to each other in the otic region of the skull. At the end of each semicircular canal is a dilatation called an ampulla which contains a sensory organ, the crista. The crista consists of sensory hair cells which project into a gelatinous partition, the cupola. The cupola closes off the ampulla like

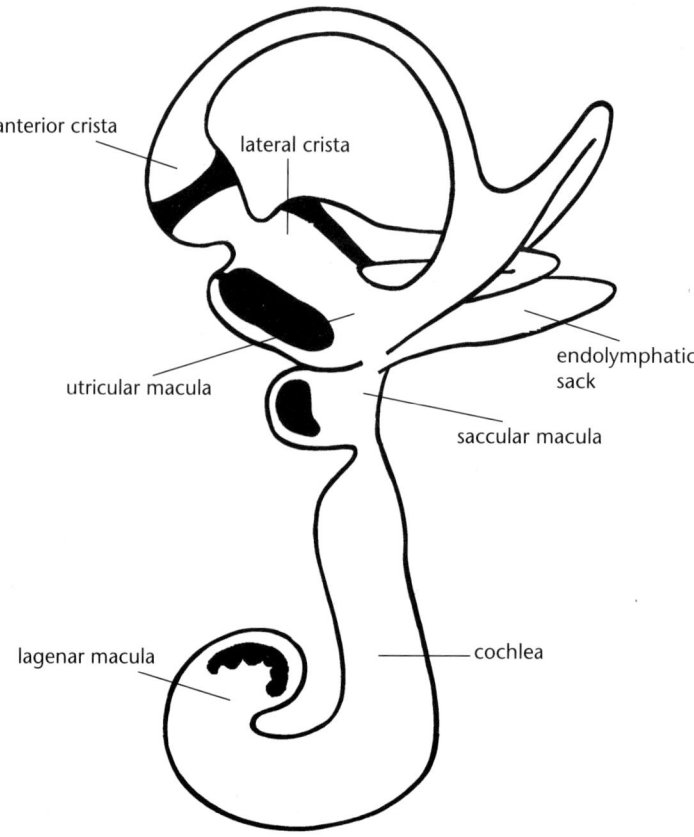

Figure 5.5 Arrangement of the vestibular organs of the right inner ear viewed medially.

a swinging door. When the head of the echidna is rotated in the plane of a particular semicircular canal, the fluid in the canal is displaced in a direction opposite to the direction of head rotation. This fluid swings the crista thereby stimulating its hair cells, producing an increased rate of neural activity. In this way the semicircular canals feed information to the brain about rotary movements of the head in three dimensions.

The utricle and saccule each possess a sensory structure called a macula. The macula consists of a basal layer of hair cells, the hairs of which also project into an upper gelatinous layer, the otolith membrane, which possesses numerous crystals of calcium carbonate partly embedded in its surface, the otoliths. The pull of gravity displaces the otoliths and the membrane, the hair cells are distorted and the nerve fibres lying between the cells are stimulated in this way. These organs provide the brain with information about the position of the head in relation to the gravitational field. The inner ear of the echidna is large for a mammal. The area of the utricular macula is 4.5 mm^2 whereas the comparable area in humans is 4.29 mm^2 and only 0.54 mm^2 in the guinea pig. The number of hair cells in the utricular macula is the largest reported for any mammalian species, 56 300. Other mammalian figures are: humans 33 100, squirrel monkeys 11 360, and guinea pigs 9260.

Furthermore, the echidna and platypus have more vestibular organs in the inner ear than any other mammal. They possess an additional macula called the lagenar macula. In the echidna this macula consists of a narrow sensory strip at the end of the cochlea. The lagenar macula is an ancestral or plesiomorphic feature found in all non-mammalian vertebrates.

Why then does the echidna possess such a richness of gravity receptors? The utricular macula is essential for the righting reflex which allows overturned animals to quickly reposition their head and then their body upwards in relation to the gravitational field. Perhaps the echidna's need to burrow to escape danger and to obtain food has influenced this sensory facility. Echidnas spend long periods in dark burrows of their own making or complex burrows of other animals. Knowing which way is up is essential for its survival, but we really don't know the particular significance in echidnas. It remains a mystery for future research.

The snout

The snout of an echidna is a most amazing structure. It is sometimes mistakenly referred to as a beak or a nose. Its structure and function are much more complex than either of the latter. Its skeleton is formed by prolongations of the upper and lower jaw bones. In an adult animal of 4 to 5 kg with a total body length of 450 mm, the snout is about 75 mm long, that is, about 17 per cent of its total body length. The skin is black and has a slightly moist, leathery feel. The outer layer of the skin is heavily keratinised which is said to be reminiscent of reptilian skin. Towards the tip it becomes smoother and may appear shiny. It plays some critical roles in the life of an echidna, acting as:

- a sensory organ
- a mechanical probe, and
- a water conserving device

The echidna's snout is an extremely important sensory organ. Wary of something approaching or when investigating a strange object, an echidna will raise the head and sniff the air. When foraging for food, it will sniff or sometimes snuffle through mucous secretion around the nostrils producing clusters of small bubbles. During foraging the echidna ambles along poking its snout here and there on or into the ground, under rocks or into decaying logs. Mervyn Griffiths described the behaviour thus:

> '... suddenly it will dart from the line of march and dig rapidly removing stones if necessary with claws and snout and will lick up the exposed ants or termites as the case may be; the whole performance has an air of more than a little ferocity.'

These behavioural observations suggest that the sense of smell plays an important part in the echidna's defensive and prey locating abilities.

Smell

The snout acts as a conduit for air to bathe an enormous area of olfactory epithelium. The olfactory epithelium is carried on vertical bony septa called ethmoturbinals (Figure 5.6) which arise from the ethmoidal bone. These ethmoturbinals originate at the cribriform plate which is pierced by a large number

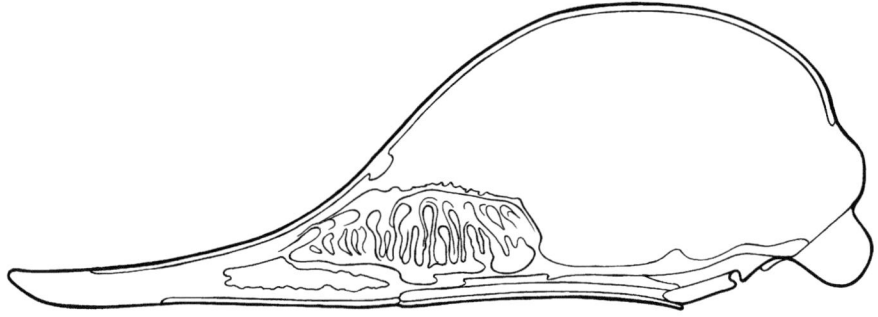

Figure 5.6 Sagittal section of the skull of an echidna showing the nasal cavity with olfactory surfaces of the vertical bony septum (ethmoturbinal).

of pores through which pass numerous branches of the olfactory nerve to the olfactory bulb. The echidna has 13 times more olfactory nerve fibres than the platypus, suggesting the relatively greater importance of the sense of smell in the echidna. The organisation of the olfactory system in the brain follows the usual mammalian pattern but there are some variations. The end-station of sensory nerve from the skin of the snout (the trigeminal nerve) is in the part of the brain called the thalamus, and the end-station is connected to the olfactory system, linking tactile sensibility of the snout with smell.

Regrettably, no physiological or controlled behavioural studies of olfaction have been carried out on echidnas. Nonetheless, anatomical and observational evidence suggests that its sense of smell is important to the echidna in at least three ways.

- Echidnas are solitary animals for most of the year but olfaction may be important in bringing the sexes together in the breeding season. At this time males find and follow females, which strongly suggests the action of a sexual pheromone, but controlled studies have yet to be performed.
- Olfaction may play a part in detection of ants and termites but no specific chemical sensitivities have been demonstrated. Lack of sensitivity to formic acid has been reported.
- Sense of smell may be important to the hatchling in finding the milk patch since its sight and hearing are not fully developed at that stage.

In the 1950s, Ludek Dobroruka made some interesting observations over a six-year period on two echidnas housed at the Prague Zoo. Initially the echidnas responded to him by rolling up in typical defensive fashion with repeated snorting.

> 'Afterwards my clothes, which to us smelled unpleasantly of echidna, smelled however to the anteaters so familiar that I could quietly stroke them without their rolling up.'

Dobroruka also provided photographic evidence of what he believed to be marking behaviour. The animal turned its cloaca inside out and wiped it on the ground leaving a glossy covering that smelt strongly of echidna. Apparently this behaviour had not been reported previously. Whether it occurs in the wild in unknown. A number of different types of glands open into the cloacal region but little is known of the composition and physiological function of the glandular secretions.

Sensory receptors in the snout

Electrophysiological recordings from fine filaments of branches of the sensory nerve from the snout (the trigeminal nerve) indicate that the echidna's snout is richly endowed with mechanoreceptors. Some of these receptors have very low thresholds. The presence of this large population of sense organs offers an explanation for the lavish supply of branches from the huge trigeminal nerves. The number of openings in the nasal bones for the peripheral distribution of the nerve is correspondingly large. Given the multi-functional role of the echidna's snout it is not surprising to find that it is well endowed with sensory endings which can be described largely under two headings: electroreceptors and mechanoreceptors (Figure 5.7).

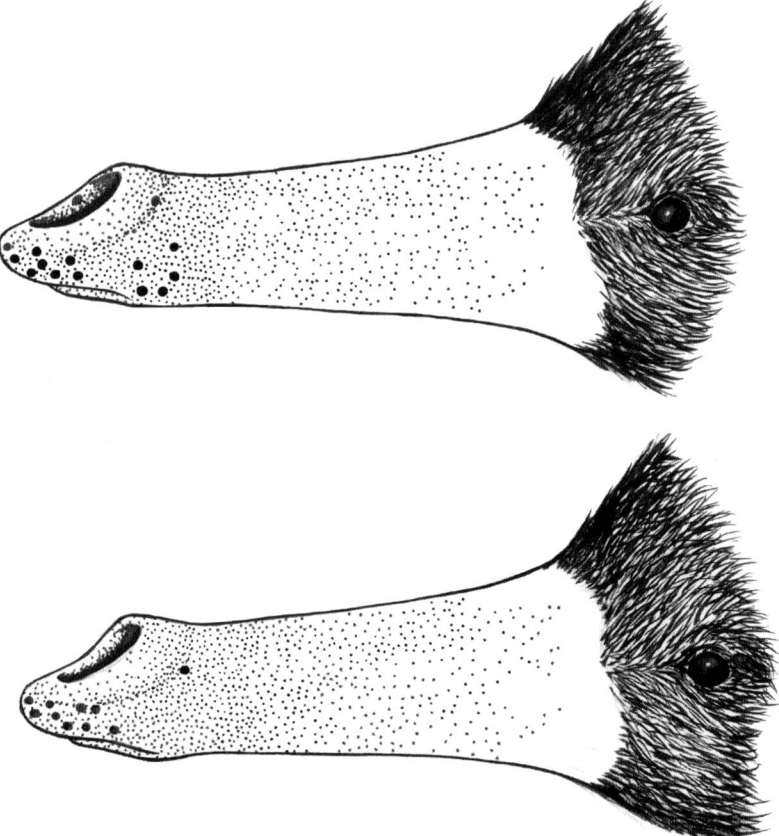

Figure 5.7 Distribution of mechanoreceptors and electroreceptors on the snout of an echidna. The upper view shows the location of the mucous secretory glands (large black dots) determined by microscope study of skin sections. The lower view shows the location of 10 electrosensitive spots determined by weak electrical stimulation of the skin surface. The fine stippling in both views indicates the location of structures in the skin that are thought to be touch receptors.

(From Proske 1990. Reproduction with permission from *Australian Natural History*).

Electroreceptors

Platypuses can find 1.5 volt miniature alkaline batteries placed at the bottom of a pool – presumably by detecting the direct current field they produce in the water. Research has shown that the platypus snout, with its mucous secreting glands with associated nerve terminals, is the electrosensitive organ. The echidna has similar receptors, albeit in a simpler form, and can detect weak electric fields – down to a strength of 1.8 mV cm^{-1} – but the receptive field is restricted to the tip of its snout.

Humans are readily able to detect the voltage of a 1.5 volt battery with the tip of the tongue but both platypuses and echidnas respond readily to voltages 1000 times smaller and far below the detection level of human beings. The response can be evoked by both steady and alternating electrical potentials.

The echidna has about 400 mucous glands in the region surrounding the snout tip and the nostrils, and a quarter of these contain sensory nerve terminals. The tip of the snout is quite small so this actually represents a density of about seven mucous sensory glands per square millimetre in the region of the lips. Each sensory gland is surrounded by several non-innervated glands in a regular arrangement suggesting some functional relationship.

The sensory mucous gland has a coiled secretory portion in the lowest (subdermal) region of the skin and the gland duct ascends to

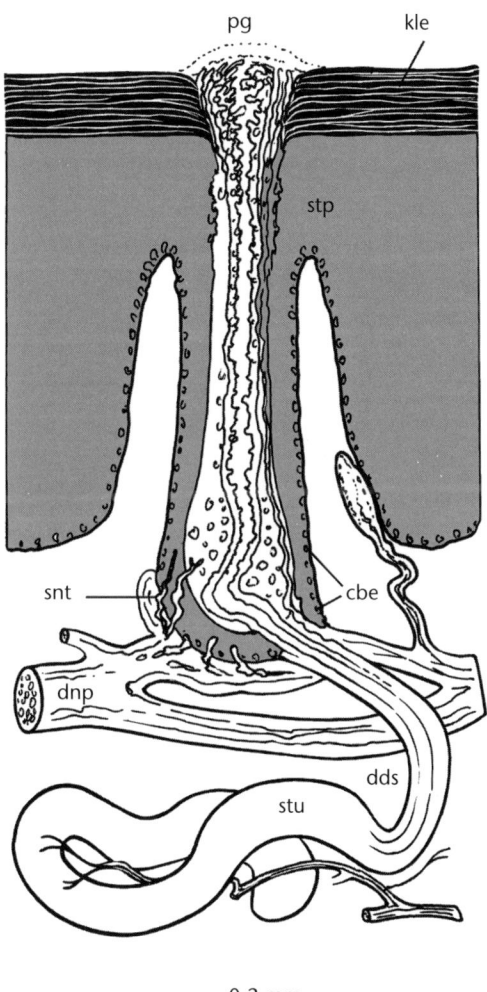

Figure 5.8 Structure of a seromucous gland with sensory innervation of the duct (an electroreceptor). Abbreviations:
pg = pore of gland, kle = keratinized layer of epithelium, stp = straight segment of the duct, snt = sensory nerve terminals adjacent to the duct, cbe = cube shaped enlargement of the epithelial papilla, dnp = dermal nerve fibre plexus, dds = dermal duct segment, stu = secretory tubule.

the skin's surface (Figure 5.8). The portion of the gland within the outer layer of the skin (the epidermis) has a club-shaped enlargement which is formed by an invagination of the epidermis. Sensory nerve terminals arising from an adjacent dermal nerve network penetrate the epidermal layer and terminate adjacent to cells clustered in the club shaped enlargement. Two kinds of nerve terminal have been identified though their specific functions are unknown. It is not known how the small electrical currents excite the nerve terminals. There are innervated epidermal pits with no apparent associated glands which may also serve an electroreceptive function. These structures are distributed just behind and lateral to the nostrils, an area with electrosensitive sites.

The functional significance of electroreceptors in the echidna is still uncertain. In a field study we found that echidnas dug up live 9 volt batteries buried in their path significantly more often than flat batteries. A totally blind echidna, who had lacked eyeballs from birth, produced the most spectacular uncovering of live batteries observed in this study. It was as though this animal had become particularly attuned to this sensory modality. It appears therefore that the electroreceptive function can be utilised in a natural environment (see Chapter 8).

The structure of the sensory mucous gland receptors in *Zaglossus* is intermediate between those of platypuses and *Tachyglossus*. Unlike platypus and *Tachyglossus*, all the mucous glands appear to be innervated. The densities of the electroreceptors are: platypus 30 per mm^2, *Zaglossus* 12 per mm^2 and *Tachyglossus* 7 per mm^2. Perhaps these values tell us something about the relative importance of electroreception in these three monotremes and their evolutionary adaptation to different environments. The main prey species for *Zaglossus* is earthworms which inhabit the floor of the humid montane rainforests. Perhaps the movement of earthworms in the moist earth produces an electrical field detectable by *Zaglossus*?

Push rods

Apart from the electroreceptor, there is another unique structure in the skin of the platypus bill and the echidna snout called the push rod. This structure was found initially in the platypus and more recently identified in *Tachyglossus* and *Zaglossus*. The push rod is distributed uniformly across the entire surface of the echidna snout and is somewhat less differentiated than that of the platypus.

The push rod is composed of a column of flattened, spinous cells which are cross-linked by junctions between adjacent cell membranes known as desmosomes (Figure 5.9). The rod is differentiated into a central and peripheral compartment. The concentration of the desmosomes is more prominent in the central than in the peripheral compartment. The size of the push rods varies at different skin regions but it is typically 300 µm long with a diameter of 50 µm.

Senses 71

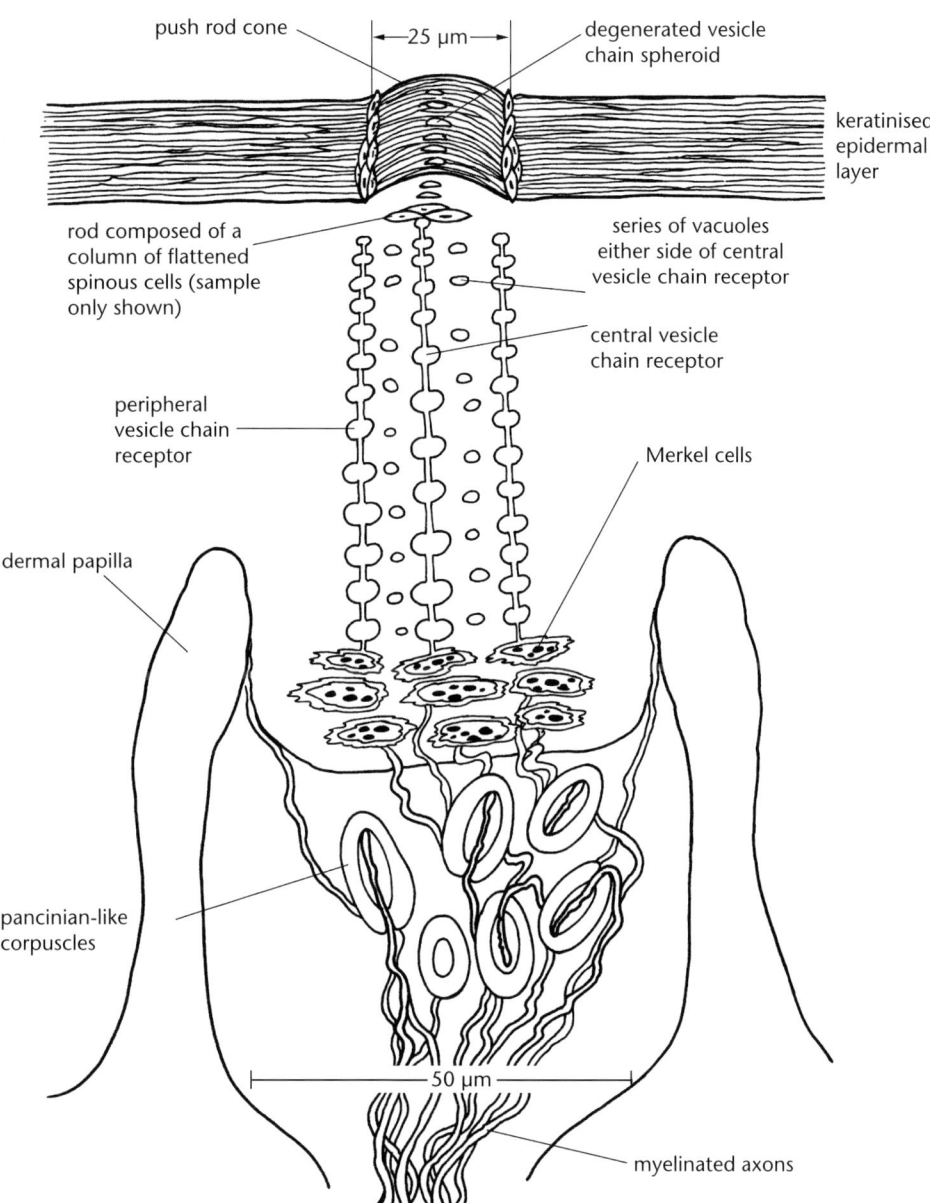

Figure 5.9 Structure of a push rod.

In a good light the dome-shaped tip of the rod can be seen on the skin surface and is about 25 μm across. This projection is called the push rod cone.

There may be as many as 30–40 push rods per mm² of skin. Along the lower two-thirds of its length, the push rod is separated from adjacent epithelium by a protrusion of connective tissue of much more loosely packed cells called a dermal papilla. It is believed that this arrangement gives the rod the capacity for a degree of independent movement.

There are three kinds of nerve endings associated with the push rods: vesicle chain receptors, Merkel cell receptors and Pancinian corpuscles. This remarkable complex of nerve endings is innervated by a total of approximately 10 myelinated axons per push rod which arise from a nearby dermal plexus. Vesicle chain receptors are found in both the central and peripheral compartments of the push rod. They number between 10 and 24 depending on the size of the rod. A vesicle chain receptor is an unmyelinated axon which passes up the length of the epidermal push rod. The axon possesses a series of flattened spheroidal dilatations which gives the appearance of a chain of pearls. The central vesicle chain receptors are surrounded by two layers of cells known as keratinocytes and these cells in turn are surrounded in parallel by a number of peripheral chain receptors. A series of vacuoles in the cytoplasm of keratinocytes runs in parallel between the central and peripheral vesicle chain receptors. At the base of the rod, 18 to 26 Merkel cells lie in up to three layers and are innervated by about four axons. In the dermis below the rod are found five to eight Pancinian corpuscles supplied by two to three axons.

The skin of the echidna's snout has to be tough enough to withstand the trauma imposed on it when it is used as a digging tool in hard ground or as a lever to break open rotting timber. On the other hand, the echidna relies on the subtle sensory perception of its snout to detect its prey. The thickened cornified layer of skin on the snout would therefore appear to be less than optimum for the transfer of the necessary sensory information. As early as 1885, E.B. Poulton suggested a possible function of the rods:

> 'The obvious use of the rods is to supply special movable areas
> (of skin) yielding to surface pressure which is thus communicated
> to the terminal organs below'.

If Poulton is right, the push rods may be thought of as channels through the tough outer layer of snout skin for the passage of mechanical stimuli arising in the external environment to the underlying sensory receptors. Can we make any sense of the sophisticated sensory arrangement of the push rods?

The application of mechanical stimuli to the snout skin using fine probes vibrated over a range of frequencies results in recordable traffic along fine strands of the infraorbital nerve, which supplies the skin of the upper jaw. Some receptors adapt slowly to this stimulus in the range of 100 to 500 Hz and are believed to be of the Merkel cell or Ruffini ending type. Other receptors adapted rapidly to the stimulus and were very sensitive to vibrations in the range of 200 to 600 Hz. These receptors are presumed to be Pancinian corpuscles. With regard to the vesicle chain receptors in the push rods, it has been suggested that the spheroids in these structures appear similar to discoid receptors noted in the skin of other animals where they are believed it is involved in the sense of touch. The vesicle chain receptor could be a specialised form of discoid receptor. The central and peripheral arrangement of these receptors may provide a system for detecting the direction of the incoming stimulus on the dome-shaped tip of the push rod. This function could be facilitated by an increase in tissue flexibility due to the string of vacuoles and decreased density of desmosomes in the keratinocytes lying between the central and peripheral chains thereby facilitating differential movement of the core and outer chains.

Despite what appears to be an obvious association between push rods and mechano-sensory function, attempts to prove this experimentally have been tantalising but frustrating. Studies designed to correlate the stimulus of a push rod with nerve impulses coming from the receptors associated with that rod have not been conclusive. Recording from receptors within a push rod when that rod is stimulated will be required to finally establish the role of push rods.

Some supporting evidence comes from a consideration of sensory receptors in other animals. For example placental moles possess a structure called the Eimer's organ which is very similar to a push rod. Numbering in the thousands, they are in the skin of the tip of the mole's elongated snout. This latter structure also has a column of epidermal cells containing a cluster of bare nerve endings and several Merkel receptors. Interestingly, there is a single axial terminal surrounded by a series of satellite terminals arranged in circumferential array (as compared to the arrangement of vesicle chain receptors in the echidna's push rod). In the dermis deep to each peg is a lamellated end organ. Eimer's organs have also been described in the star-nosed mole, *Condylura cristata*, whose star-shaped nose is used to probe mud and soil for prey items such as beetles and worms. The skin of the star contains densely packed Eimer's organs but few other receptors. Recordings made from the cerebral cortex revealed discharges in response to mechanical stimulation of the star. Since Eimer's organs resemble push rods, these findings in the star-nosed mole are supporting evidence that push rods are mechanoreceptors.

Non-encapsulated Ruffini endings are abundant in the connective tissue joining the bones of the snout skeleton and their stretching may play a part in protecting the snout from serious injury such as fracture. In addition, numerous free nerve terminals in bone cavities may signal to the brain by means of pain perception that structural tolerance limits are being approached.

We are just beginning to understand how complex and subtle the echidna's snout is for the detection, identification and final consumption of prey. But we may be at a point where we can begin to see how it operates as a sophisticated integrated system. When the snout touches the ground it can apparently pick up mid-frequency sounds around 5000 Hz emanating from ant and termite colonies, possibly some distance away, by bone conduction to the inner ears. The anatomy of the external ear canal is designed to damp down confounding frequencies which might come by air conduction. This snout bone conduction of sound could be considered as the echidna's long-range prey detection system. Prey such as beetles and larvae may vibrate their wings or bodies against or in the soil, especially when trying to escape, producing vibrations in the 500 Hz range; that is, an order of magnitude less than ant and termite colony frequencies. Of course, vibrations in solids which produce compressional waves are usually considered to be sound waves, so we may consider this kind of prey produces low frequency sound. The ear is not specially attuned to this range of frequencies – this low frequency sound is presumably picked up by the rod receptors. The varying strength of signals from these receptors arranged along the length of the snout may also give directional information. Further information about the prey type and distance may then come from the reaction of electroreceptors to the electrical field in the soil produced by the muscular activity of the fleeing victim. Variation in the field strength registered by the electroreceptors distributed across the tip of the snout may also indicate the direction of the prey.

Finally the snout closes on the prey and contact is made. The prey touches the skin of the snout, the domes of the rod receptors come into contact with the prey and the specific point of contact on the dome gives precise information about its exact location. In addition, these receptors are so sophisticated that they may also give details about the texture of the surface of the prey, which could identify the part of the prey's body that has been contacted. This fact could be important if the prey has some noxious defence strategy. The prey can now be manipulated into the mouth or broken down to an appropriate size for ingestion.

Venous cavernous system

Another intriguing aspect of the snout is that it possesses an extensive system of blood filled cavities, the venous cavernous system, lying between the bony snout and the skin in the dermal and sub-dermal connective tissue. Engorgement of this venous tissue would be expected to increase the size of the snout, but to what purpose? It has been suggested that the consequent outward bulging of the skin, if activated during a snout prod, would make better contact with the soil and hence raise the sensitivity of the skin sensory receptors. This concept was proposed some years ago in relation to the Eimer's organs of the mole.

Two other possible functions for this venous cavernous system may be postulated. When sound crosses from one medium to another, partial reflection occurs and there is a consequent reduction in the transmitted sound intensity. Thus a fluid layer between the bone and the skin introduces two discontinuities in the acoustic path, which may help to reduce 'cross-talk' between the higher frequency sound passing up the snout by bone conduction and lower frequency sound entering the sides of the snout through the skin.

To get its snout caught in a tight place could mean serious injury or even death to an echidna. One way to reduce this risk might be called the 'Houdini effect'. The famous escapologist Harry Houdini contracted his muscles during the application of ropes, thereby making his limbs effectively larger than they were normally. Subsequent relaxation of the muscles allowed for easier removal of the bonds. In the echidna, the most prominent region of the venous cavernous plexus lies on the dorsum of the snout, just behind the nostrils. This location has been called the postnasal bulge thickening. It is possible that the echidna activates engorgement of the snout prior to its insertion into locations which might represent a threat of entrapment. The postnasal bulge (shown around the nostrils in Figure 5.7) when engorged would be like a feeler gauge acting as a safety monitor in relation to the size and tightness of the hole into which the rest of the snout is being inserted. If the engorged snout could fit into a hole, then, *ipso facto*, the flaccid snout should be removed easily. The evolutionary pressure for the development of such a safety device for the snout is obvious.

6
Reproduction

Since the echidna is such a secretive animal, it is not surprising that much of its reproductive behaviour has remained a mystery for so long, despite the best efforts of scientific investigators. Early in the nineteenth century the question of whether echidnas laid eggs or gave live birth was a hotly debated topic. The concept of a mammal that laid eggs was novel, but that question was settled long ago and monotreme oviparity firmly established. However, many other aspects of echidna reproduction remain uncertain even today.

In this chapter we will explore the life cycle of the echidna, from courting to independent young.

Courtship

Echidnas are not usually gregarious. They move about their home ranges independently, although these ranges may overlap with those of other echidnas of both sexes. They do of course come together at mating time, which is now generally agreed to be from June to early September (Figure 6.1). This period appears to be the same regardless of geographical locations as diverse as south-east Queensland, Kangaroo Island, South Australia, Kosciuszko National Park and Tasmania.

The initial act of courtship is pursuit. Groups of as many as 11 echidnas have been reported during the mating season in various parts of Australia. The

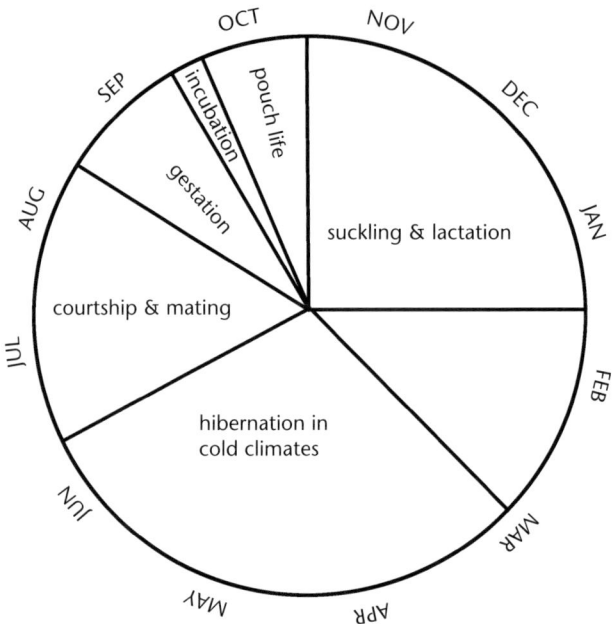

Figure 6.1 The annual cycle of an adult female echidna with young. The point at which the young is left behind in the nursery burrow while the mother forages (the end of 'pouch life' in the diagram) is uncertain and probably variable.

animals are arrayed in 'trains' in which the female is at the head and the males follow in 'Indian file' nose to tail. These trains are dynamic in their composition. The number and identity of the males vary from day to day. A study of Kangaroo Island echidnas found that the duration of courtship, defined as the time from when a female was first observed with males until she mated, ranged from 7 to 37 days. These groups of echidnas forage, walk and rest together during this courtship phase. The smallest and often the youngest animal is usually at the end of the train. In the Snowy Mountains, however, no trains have been seen to date. Pairs form and mate almost immediately after arousal from hibernation. The severity of the climate may explain this difference in reproductive behaviour.

How are the sexes in these usually solitary animals brought together? Over a hundred years ago Richard Semon noted pithily:

> 'Particularly during the rut, both sexes produce a most
> conspicuous odour, which is probably destined to favour the
> mutual approach of the animals and enhance sexual excitement.
> This very odour gives the meat of the animal, when roasted
> within its skin, its particular and in my opinion, very nasty taste.'

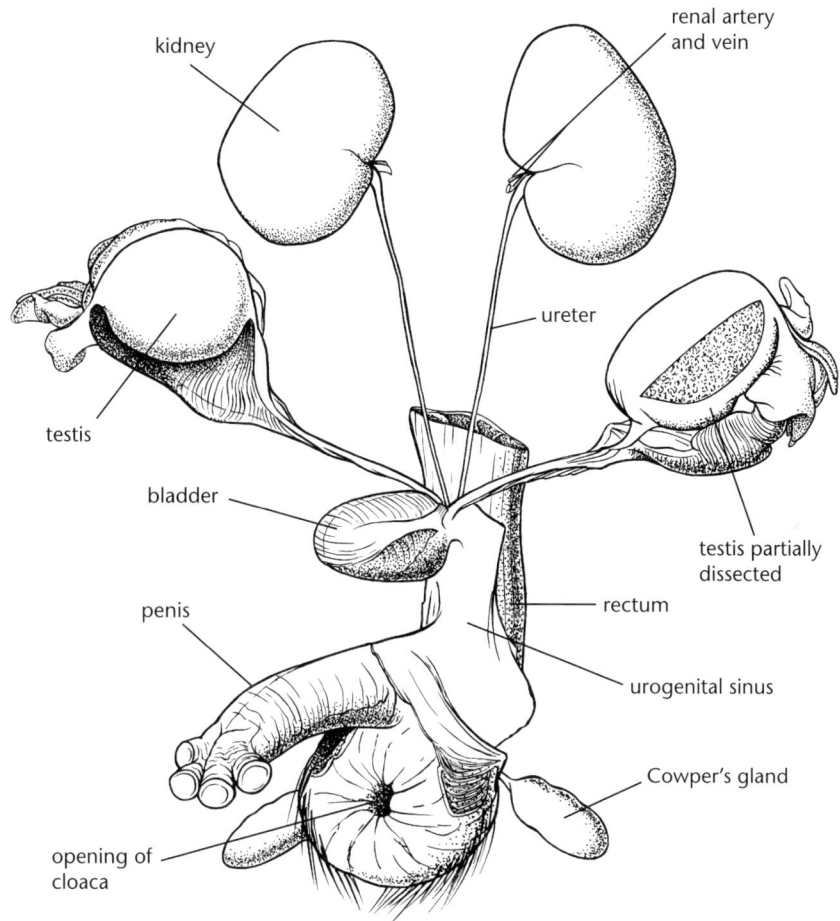

Figure 6.2 Male reproductive tract of the echidna.

It seems very likely that echidnas are brought together by following odour tracks. A captured echidna of unspecified sex was photographed turning its cloaca inside out and wiping a glossy secretion, which had a strong 'echidna' smell, on the ground. A specific pheromone may be produced during the mating season. It has been observed on Kangaroo Island in South Australia that during the mating season male echidnas were attracted to a hessian bag which had previously held a female echidna. A strong musky odour has been observed coming from the direction of a female in a train of echidnas at rest. Similarly a musky-smelling, oily, clear, yellow secretion was noted coming from a female echidna which was found in her hibernaculum in the Snowy Mountains in August. This secretion may have been a sexual attractant.

Male reproductive tract

The unusual reproductive tract of the male echidna is often overlooked but deserves description. In monotremes, the two ovoid testes are suspended within the abdominal cavity (an arrangement referred to as testicond) behind the kidneys (Figure 6.2). The intra-abdominal site, rather than scrotal, is not however unique to the monotremes; it also occurs in some eutherian mammals such as elephants and whales. Monotremes have relatively heavier testes than marsupials, primates and avian species, although the reason for this is unknown. Tubules in the testis, which are the site of sperm formation, pass to the anterior end of the testis where they form a number of ducts which continue to the epididymis. A short vas deferens conveys the sperm from the terminal epididymis to the urogenital sinus. Sperm can pass down this sinus and then enter the urethra of the penis. The urethra, despite it name, passes only sperm, not urine. There are no seminal vesicles and no discrete prostate gland.

In the echidna the development of sperm – that is, spermatogenesis – is seasonal. From October to March the testes weigh 1 to 3 gm/kg of body weight. In early April this increases up to 8 gm/kg and in April/May spermatogenesis begins. The highest testicular weights are recorded in August and September. A rapid regression follows and, by the end of September, the testes have shrunk back to bean-sized organs.

The structure of monotreme sperm is distinctive and its simplicity is unrivalled amongst mammals. The sperm released from the testes are immotile but further maturation in the epididymis activates their flagella. The terminal segment of the epididymis is structurally quite different from other mammals and appears to have a secretory function. The structure of the epididymis, the sperm and its maturation there, all resemble more closely these aspects in birds than in marsupials and eutherian mammals.

Mating

The precursor to intercourse appears to be the prodding of the female's body and cloaca by one or more males, as well as sniffing along her back or side from head to tail. This process, which may last for several hours, suggests the possibility that receptivity in the female may be signalled by a particular odour. If the female is not receptive she will curl up to form a spiny ball thereby rebuffing the one or more males.

The first observation of echidnas mating is probably that reported in 1895 by Robert Broom.

> 'Mr. Angus McInnis … states that the two [echidnas] were lying together on a slight hollow at the root of a tree, and so far as he

could observe front to front but on his near approach the two separated and endeavoured to escape … on picking up the male its copulatory organ was protruding about a couple of inches; so that there can be little or no doubt but coitus had just taken place.'

In captive animals, the most frequently observed mating position involved the male rolling the female onto her side and assuming a similar position so that the approach was abdomen to abdomen. In the wild, mating behaviour is more complex. During field studies on Kangaroo Island, Peggy Rismiller and her colleagues found that the female apparently signals receptivity by lying flat on her stomach, often with her head and her front feet dug in at the base of a shrub or small tree. The male or males dig along the side of the female. If more than one male is present, they push each other with their heads until one male is dominant. This male then continues to dig away the earth under the female's tail until he can use his hindlimb to lift her tail. He then lies on his side in the trench and inserts his extended penis into her cloaca. Lying in a trench may also support the male and help prevent him from rolling over onto his back. Intercourse is said to last for 30 to 180 minutes. The animals then go their separate ways. Copulation can take place in either of two positions, head-to-head or heads facing in opposite directions. In both cases tails are hooked together with the male lying on his side and the female flat on her abdomen.

The penis is a truly remarkable structure being about 7 cm long when protruding from the cloacal opening. The tip of the penis (the glans) is split by a groove as it is in marsupials. Each half has two small bulbous processes each of

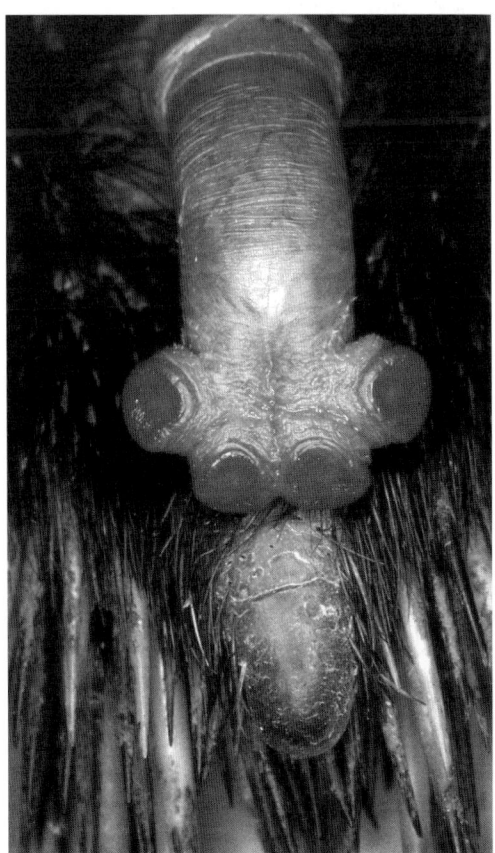

The echidna penis is divided into two halves at the tip, and each half has two bulb-like knobs, here engorged with blood. Photo: Gordon Grigg.

which bears a rosette of epidermal rays which have a striking red colour. Urine passes from the bladder via the urogenital sinus directly into the cloaca in both sexes. The urethra splits into a branch entering each half of the tip of the penis and splits again into the two bulbs of these halves. When not erect the penis is retracted into a ventral sac inside the cloaca. The bifid nature of the glans may assist injection of sperm into the paired uteri which open separately into the urogenital sinus.

Females usually mate only once in a season. However, Lyn Beard and Gordon Grigg found that in south-east Queensland females could conceive successfully a second time within the one season if the first young was lost. Within 48 hours of mating, females have been observed in the field to return to a solitary life, while males either joined another train or returned to their home ranges. These observations suggest that the pheromone attractant ceased to be excreted by the female after copulation. Further research on the nature of echidna pheromones would be of considerable interest.

Over 100 years ago George Bennett inferred from the examination of adult female echidnas that they only reproduced every second year. The reality is more complex. The work of Peggy Rismiller and Mike McKelvey has revealed that although females can reproduce in consecutive years, this pattern is unusual. The frequency of reproduction is individualistic and variable.

The egg

Fertilisation takes place in the oviduct where the initial phases of deposition of the eggshell take place (Figure 6.3). In the uterus itself, various secretions are added until the final tough outer coat of the shell is applied. The shell is not mineralised and is leathery in nature when laid. Usually only one egg is impregnated and developed at a time – is is rare to find a female carrying two young in her pouch.

Gestation

Little is known of the behaviour of the female during the first weeks after fertilisation. It is probable that at this time she constructs an incubation or nursery burrow or refurbishes one she has used in previous seasons. Breeding females observed in the Kosciuszko National Park entered a specially dug nursery burrow shortly before the time when they would have expected to have laid the egg.

The location of the burrow varies according to geography and opportunity. Some sites may offer easy digging – such as in piles of soil intended for roadworks or in the base of red soil termite mounds. Other sites require considerably more effort such as hard soil between the roots of a tree or, above the

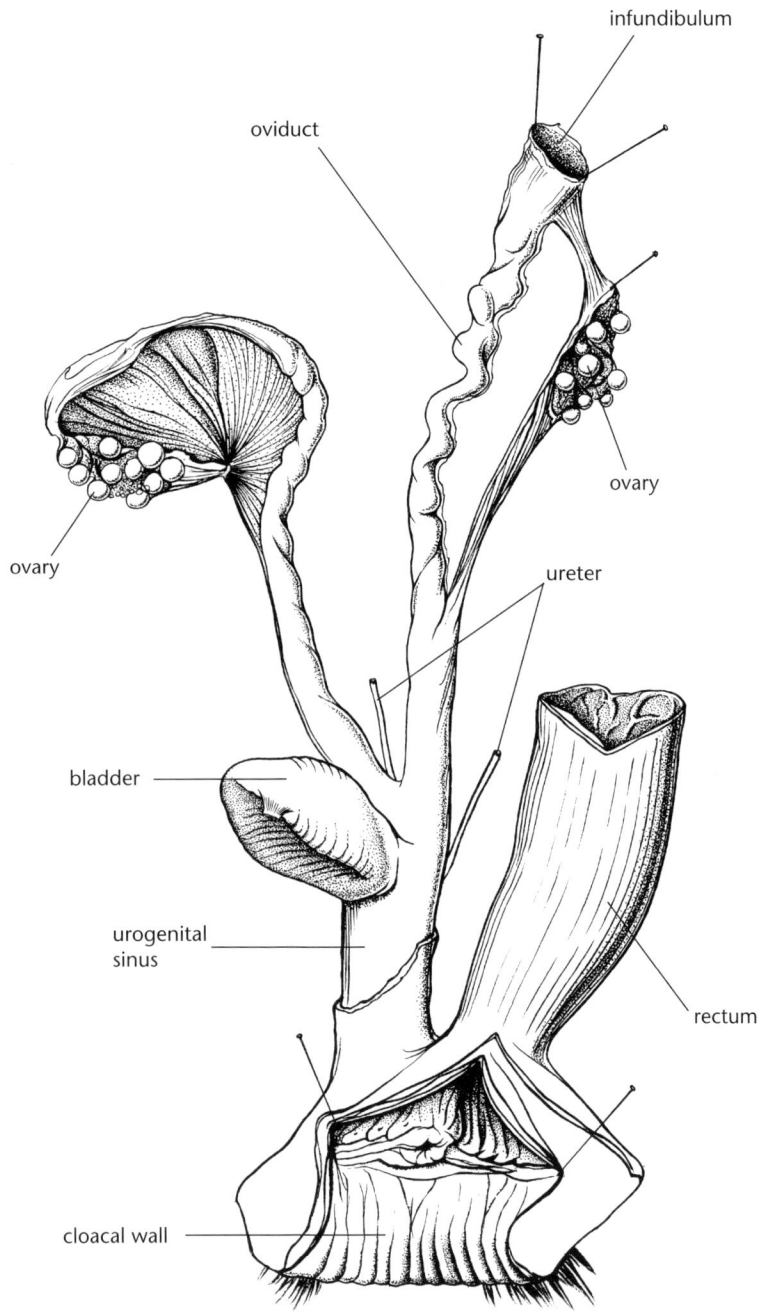

Figure 6.3 The female reproductive tract, bladder, ureters, rectum and cloaca of the echidna. The cloaca is cut and pinned open to display the opening of the rectum and the opening of the urogenital sinus. The ovaries show large follicles relative to the follicles of most female mammals. The right ovary has been separated from the oviduct and the opening (infundibulum) displayed. As shown, the ureters do not connect directly to the bladder but empty into the urogenital sinus.

snowline, in the base of a mound of matted vegetation. In the Kosciuszko National Park, most burrows consisted of a tunnel running about one metre, ending in an enlarged chamber (see photograph on page 31). One burrow was reported in 'an utterly alien environment' – a mound of garden refuse, fine black earth, various weeds and blackberry plants situated a short distance from a busy road, workshops and offices. On Kangaroo Island, Peggy Rismiller found that burrows were either entirely self-dug or excavations in natural rock or tree root crevices.

In the 1890s, Richard Semon claimed that during the breeding season, whether conception occurs or not, the female develops a single pouch at the midline of the abdominal wall. However, this development is much more pronounced if the female is pregnant. The ventral abdominal region of both sexes lacks spines and has a central concavity where hairs are sparse. Ridges, formed by abdominal muscles when an echidna curls its body while lying on its back, give the appearance of a pouch in males as well as females. This region becomes a true pouch in the female as these ridges thicken. In pregnancy this hypertrophy is believed to involve all tissues including muscle, not just mammary tissue.

Egg laying and incubation

In her field studies on Kangaroo Island, Peggy Rismiller regularly checked the pouches of 10 females who had mated, and found an egg in the pouch 22 to 24 days after copulation. Egg laying has only very recently been observed for the first time. Rismiller and McKelvey describe the event as follows:

> 'When female YGBL was located on 28 July 1994, she was in
> a sitting position with the tail curled up towards the pouch.
> She extended the cloaca above the anterior edge of the pouch
> obscuring the opening. The cloaca contracted and was retracted
> into the body. After <30 s, she uncurled and walked away. We
> retrieved her and found a very white egg in the pouch.'

The cream-coloured egg is soft and leathery, with a round to slightly oval shape. It ranges in size from 13 to 17 mm in diameter – about the size of a grape. Its weight varies from about 1.5 to 2 g. The incubation period in the pouch lasts about 10 days. In the Snowy Mountains the female enters her nursery burrow shortly before laying the egg, but on Kangaroo Island some females continue to forage on the surface while others dig a burrow and remain there until the egg hatches. In south-east Queensland, Lyn Beard and Gordon Grigg found that females spent two to three weeks in the incubation burrow.

Hatching and pouch life

At an early stage in the development of the embryo, the egg tooth begins to form near the junction (symphysis) of the front bones of the upper jaw (premaxilla). At first it is a simple midline papilla, but soon a layer of dentine is formed as well as an enamel covering. When development is completed during incubation, the unpaired egg tooth has the appearance of a sharp incisor pointing downwards over the mouth (see Figure 3.2). It is used to tear open the eggshell at hatching and is aided in this action by the 'caruncle', which is a pointed bump on the end of the snout composed of hardened epithelium. Both the egg tooth and the caruncle disappear soon after hatching. The hatchling is only 13–15 mm from head to tail when it works its way out of the shell. Although its eyes are rudimentary, its front limbs are well formed with digits and miniature claws, allowing it to pull itself along the pouch hairs. The hatchling weighs only 0.3 to 0.4 g at birth. Its body is semitranslucent and is draped in the remnants of the yolk sac.

Lactation and suckling

The echidna's mammary tissue has no nipples but the mammary glands open onto two areas of skin in the anterior part of the pouch known as the milk patch or areola. This area contains between 100 and 150 separate pores, each

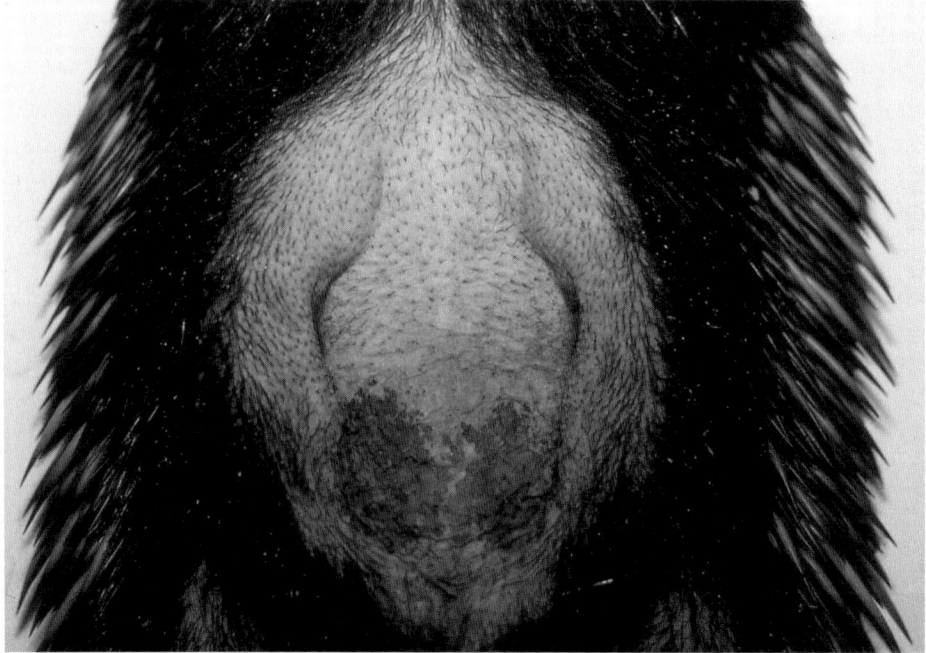

Mammary glands on the shaved underside of a female echidna. The mammary glands open onto two areas of skin known as the milk patch, or areola. Photo: Gordon Grigg.

with a specialised hair follicle. The baby echidna may use its sense of smell to direct it to the areola. It does not lick the milk from the skin as was once thought but actually sucks it from the areola.

Pouch young (see photographs, page 30) grow quickly, with their weight increasing from around half a gram up to 400 g in about 60 days. Young echidnas in captivity have been shown to ingest large quantities of milk every two to three days. At one suckling the young can increase its weight by up to 20 per cent in one to two hours. Intervals between suckling in free-living echidnas can be considerably longer – from 5 to 10 days. A female was found to forage away from the nursery borrow for this length of time. The average increase in the body weight of echidna young per ml of milk consumed has been found to be 0.4 g per ml. The rate of weight gain of the young is directly related to the body weight of the mother, that is, the young of small mothers tended to grow more slowly than those of larger dams.

While radio-tracking echidnas in south-east Queensland, Lyn Beard and Gordon Grigg found that a young was carried in the pouch for 45–50 days. When it weighed about 200 g the young was left behind in the plugged nursery burrow. The mother returned to the burrow regularly at intervals of four to six days until the independent young emerged at about five-and-a-half months of age.

Echidna milk

During his detailed studies of reproduction in the echidna, Richard Semon found the alimentary canal of the young filled with a white milk-like substance. The whitish colour he attributed to the presence of numerous fat globules. In the stomach he noted a solid clot rather than a fluid. This 'coagulum' he noted appeared to lack 'sugar of milk' – presumably a reference to lactose. Consequently, Semon believed that echidna milk differed somewhat in its chemical composition to that of 'higher mammals'.

Since that time the constituents of echidna milk have been precisely defined. The mother's milk at hatching is dilute: it consists of 1.25 per cent fat, 7.85 per cent protein, and 2.85 per cent carbohydrate and minerals. Mature milk, by comparison, is very concentrated with 31 per cent fat, 12.4 per cent protein, 2.8 per cent sugar and other components.

Unlike the milk of eutherians, echidna milk contains little free lactose. The fat content is similar to other mammals but the mature echidna milk contains very little polyunsaturated fatty acids, unlike platypus milk which contains large amounts of these fatty acids. Milk sampled nearer weaning has been found to contain even more protein than the 'mature' milk, and this may be related to a demand for keratin synthesis associated with growth of hair and spines in the young animal about to emerge from the burrow and face the approaching cooler temperatures of winter as well as the possibility of predators. The principal carbohydrates are fucosyllactose and sialyllactose, making it unique even in comparison with the platypus in which the principal carbohydrate is difucosyllactose. The echidna whey, that is, the watery part of the milk that separates from the curd when the milk coagulates, contains a large amount of the iron-binding protein transferrin.

The whey of mature echidna milk is reddish-pink and this colour can be intensified by increasing the quantity of iron bound to the transferrin. The mature milk contains 33 µg/ml, which is very high compared with the milk of eutherians and greater on average to that found in platypuses. It is believed that this characteristic of the milk is necessary because the echidna and platypus young are so small at hatching that their livers are unable to store sufficient iron to tide them over until they can forage for themselves.

It has been known for some time that echidna milk whey possesses the ability to produce lactose. In eutherian mammals the synthesis of the milk sugar, lactose, is catalysed by two proteins including alpha-lactalbumin. They act in the mammary gland and are secreted into the milk. Recently alpha-lactalbumin was found in echidna milk in very low concentration compared with other species. Its amino acid sequence resembles more closely that of the platypus than that of eutherian and marsupial alpha-lactalbumins.

Another milk protein called lysozyme, which is structurally similar to alpha-lactalbumin, was considered a possible promoter of lactose synthesis in the echidna. The amino acid sequence of echidna milk lysozyme is unique. To date, lysozyme has not been isolated from the platypus. Echidna lysozyme does not promote the synthesis of lactose, even in high concentration, indicating that lactose synthesis in the echidna occurs by the same mechanism as that found in the platypus and other mammals.

Leaving the pouch

On Kangaroo Island females were found to carry young in their pouch for 45 to 55 days, by which time the young weighed between 180 g to 260 g. These observations are in general agreement with earlier work that reported a mean duration of pouch life of 53 days. At the end of this time the echidna young start to develop spines and the mother ejects it from the pouch (see photograph, page 30). She places the youngster in a burrow which she has dug specifically for this purpose. The mother keeps the entrance to the burrow blocked and replaces this plug on leaving and re-entering the burrow when she goes out to forage. The reason for this back-filling may be twofold; it maintains an equitable environmental temperature and deters predation of the young echidna.

Even within the nursery the young may not be completely safe from predation by goannas, large snakes, feral cats, foxes and dogs.

In the very cold environment of the Snowy Mountains, females do not leave their burrows until three weeks after the hatching of the young and then they do not appear to carry the young with them, even though the young have no hair or spines at this time. This different pattern of behaviour may relate at least in part to a limitation of thermoregulatory capacity in the young.

Weaning

Until recently little was known about the process of weaning. On Kangaroo Island the time of weaning has been documented at 180 to 205 days after hatching, when the young weighed 800 to 1300 g. Similarly, earlier studies had suggested a period of around 200 days. The young were observed to be weaned between mid-January and the end of February. The events were as follows. The female was observed to return to the burrow, open the entrance and bring out the young. She then nursed it and left without backfilling the burrow. No further contact was observed between mother and her offspring. In all recorded cases the female weaned the young by not returning to the burrow and apparently avoided further contact with it. Presumably the innate inquisitive nature of the young is sufficient for it to survive.

The age of sexual maturity of echidnas, both male and female, still remains a mystery. It is also unclear how quickly the young establish their own home range.

7
Behaviour

Nest sites

Echidnas are solitary and usually well hidden from human observers. Except for burrows dug by females for use as nursery chambers, echidnas have no fixed nest or shelter sites. They simply take refuge wherever appropriate during periods of inactivity. These retreat sites almost always provide good camouflage, and even if hiding in a clump of grass or just under a pile of forest litter it is impossible to pick an echidna by sight.

In field studies of echidnas with radio-tracking transmitters we have often found ourselves with a tracking antenna pointing at some bit of litter or cover that shows no sign of an echidna. The echidna underneath is only disclosed when it is disturbed. If not disturbed, an echidna will remain absolutely still, which is excellent defence for a well camouflaged animal except against predators with a good sense of smell. Dogs, for example, have recently been used by natives of both New Guinea and Australia to hunt echidnas, which they find with ease. Barking gives away the location of the echidna, although after one or two experiences with the spines dogs will not normally hassle echidnas.

Table 7.1 shows the sites used by echidnas in the Snowy Mountains of New South Wales and near Stanthorpe, southern Queensland. Hollow logs and decaying tree stumps are much favoured by echidnas, presumably because they offer both shelter and food, being a good source of termites in particular.

Those areas of eastern Australia which are known to have high densities of *T.a. aculeatus* all have abundant fallen and dead trees, and in agricultural areas echidnas are most likely to be found along roadsides or in patches of uncleared scrub. However, from field studies in Western Australia, Max Abensperg-Traun concluded that adult echidnas preferred subterranean shelter. As shown in Table 7.1, echidnas in southern Queensland made much use of rabbit burrows and in the same study 60 per cent of their hibernacula were subterranean. Echidnas are clearly opportunistic, even making use on occasion of wombat burrows and space under buildings.

Table 7.1. Shelter sites used by echidnas in radio-tracking field studies.

	Snowy Mountains	Southern Queensland
Hollow logs	30%	37%
Hollows at base of tree	18%	–
Hollow tree stumps	6%	3%
Cavities in roots of live trees	–	3%
Rabbit burrows	7%	23%
Depressions under fallen trees	7%	21%
Under rocks	7%	–
Thick patches of undergrowth	4%	3%
Under leaf or bark litter	4%	8%
Wombat burrow	3%	–
Rock crevices	3%	3%
Underlog piles (beside cleared areas)	3%	–

The percentage of use figures in Table 7.1 do not correlate with relative occurrence of nest sites, so refuges are selected. Preference for site types changes over the seasons as sites selected for hibernation show a preference for more sheltered, insulated sites. Site selection does not depend entirely on vision since a totally blind echidna radio-tracked near Stanthorpe, Queensland had the same pattern of shelter selection as other echidnas in the field study.

Over an annual cycle some nest sites may be reused but this appears to be random.

Activity patterns

The time during which refuges are used is also opportunistic in the sense that activity is related to ambient temperature rather than to a fixed daily cycle. Echidnas are primarily crepuscular over much of their geographical range, that is, they are active in the twilight of evening and the early light of morning. In this case the refuge site used at night will almost never be the same as the one used the day before and vice versa. The activity period is shifted in relation to environmental temperature, particularly heat. Echidnas have no sweat glands and do not pant. Although they show remarkable tolerance for low body temperatures, they are vulnerable to heat stress. Therefore they must avoid high temperatures and respond to hot days by shifting their activities to night (nocturnal) and seeking refuge during the day in relatively cool microclimates such as rock crevices in arid regions. On the other hand, during winter in eastern Australia and Tasmania, they can become mainly diurnal (day active).

Home range

Echidnas are solitary but mutually tolerant. They do not have a 'territory' in the sense of an area which they defend, but rather a 'home range' within which they spend most of their life. There is a great deal of individual variation in the size and shape of the home range. In a field study of radio-tracked echidnas in the Snowy Mountains, home ranges varied from 24 to 76 ha. In a similar study near Stanthorpe, southern Queensland, they varied from 21 to 93 ha, and on Kangaroo Island from 9 ha to an unusually large 192 ha.

Some of the variation in the scientific literature depends on methods used to estimate home range and the number of observations available on which to make that estimation. There may also be a problem in defining home range. In at least two field studies individual echidnas with an established home range were observed to make a sudden move of several kilometres to establish a new home range. One individual in the Snowy Mountains did just this. Then, after many months, it returned and remained within the original home range. Are these areas counted as one home range or two? In the case of the 192 ha reported for Kangaroo Island it is likely something similar was counted as one home range since other radio-tracking studies on Kangaroo Island carried out by two separate groups of biologists did not find any home ranges of that size.

Regardless of the individual variations, and variations in technique, mean home ranges reported for echidnas throughout Australia are remarkably similar: 50 ha in southern Queensland; 65 ha on Kangaroo Island; 45 ha in the Snowy Mountains; and 65 ha in the wheatbelt country of Western Australia. Overall there appears to be no significant difference between home ranges of males and females and only a weak correlation (from the study near Stanthorpe, Queensland) indicating larger individuals might have larger home ranges.

In all field studies echidnas have been found to have overlapping home ranges with other individuals. Because of the cryptic nature of echidnas, it is impossible to be sure that all individuals within a study area or even part of a study area have been located. Therefore it is impossible to ascertain how many echidnas might share any part of a home range. In the Snowy Mountains study there was one small area which we knew was included in the home range of seven echidnas. The number of overlapping ranges will depend on population density, which will ultimately depend on the nature of the habitat. In habitat where retreat sites are limited, it is possible to find echidnas actually sharing nest sites. This has been observed most often in hot areas where the number of suitable rock crevices is limited. Up to a dozen echidnas have been found in such retreats. In captivity this mutual tolerance is readily apparent. On winter mornings at the Taronga Zoo in Sydney, a number of echidnas sharing a large enclosure would form a pile in the one spot that afforded basking in the morning sun. One study of captive echidnas in a crowded enclosure showed a hierarchy based on body size – a smaller echidna would give way to a larger one when they met. However there is no evidence for such hierarchies in natural conditions, even during mating season when several males will form a 'train' following a female prior to mating. In her study on Kangaroo Island, Peggy Rismiller found that it was not necessarily the larger male that finally achieved mating with the female. Echidnas are not the only species in which studies of behaviour in captive situations have been found to be of little relevance to life in the wild.

Swimming

Echidnas have been observed swimming in dams in hot weather and swimming across streams. It is an interesting sight because they hold their snout in an upright position and use it as a snorkel. On Kangaroo Island, Peggy Rismiller has seen echidnas swimming out into the ocean a short distance and back again without being forced to do so by predators or disturbance.

Defence

Echidnas have no active defensive or offensive weapons, although the non-functional spur on the hindleg of the male might be the remnant of one such weapon. Echidna defence systems are passive. In the first instance they are secretive – they are hard to find and hard to see. Until independence the young are hidden away in burrows, the entrance to which is plugged by the mother when leaving the burrow to forage. Adult echidnas are rarely active in broad daylight unless it is very cold when the sun is not up. Any unusual sound, detected with its acute sense of hearing, will cause an echidna to 'freeze' or take cover. When echidnas take cover in a clump of undergrowth or in a pile of forest litter, their coloration makes them perfectly camouflaged. For this reason, humans rarely see echidnas in the bush unless they have stopped for a while and a foraging echidna wanders into sight. It is then possible to watch the echidna going about its business unless movement of the observer is heard. If a possible predator approaches too closely and an echidna has not found nearby shelter, it will start to burrow or press its body to the ground so that all that is exposed is a formidable array of lance-like spikes. To protect its soft underbelly the echidna can dig straight into the earth with remarkable speed, sinking straight down. If caught on a hard surface and forced onto its back, an echidna will form a tight ball of spines. An unusual, last ditch form of defence is the ejection of a powerful stream of pungent urine. If one must pick up an echidna, point it at someone else!

Life in the slow lane

In comparison to other mammals, echidnas live in the 'slow lane'. Their reproductive rate is low, with no more than one young per year and normally only one young every two years. Echidnas have a low body temperature for a mammal, rarely rising above 33°C even when they are active, and basal metabolic rate about one third that of a placental mammal such as a dog or cat. But an echidna can live three times as long as a dog or cat. The record is 49 years achieved by an echidna in a zoo in Philadelphia, USA.

The time span of fossil monotremes is incredible, going back into the age of dinosaurs more than 100 million years ago, although it is not certain at which point in the evolution of monotremes echidnas, as such, appeared. When we look at details of echidna biology, we are not looking at a 'living fossil' that has failed to join modern mammals such as ourselves in the 'fast lane', but a form that has found a niche in the 'slow lane' that is so successful that it has remained there for millions of years.

8
Food and feeding

The prey

Travel anywhere in the Australian bush and look down. Before long you will see ants. Nearby, although you will not see them in daylight, there will be termites. The known number of species of ants in Australia is about 3000 and of termites, about 350. Ant nests may be completely underground or partly above it in earthen mounds, as in the case of meat ants. About 80 per cent of termites build their nests completely underground; the remainder either build mounds (such as the magnetic and spinifex termites) or live in dead wood such as fallen tree trunks. Magnetic termite mounds may house up to 200 000 individuals while in nests of other species there may be more than a million. The nest of *Mastotermes* species can hold up to several kilograms of termites.

Echidnas are adapted to take advantage of these concentrated food sources. Their powerful, spatulate forepaws with stout claws are adapted to obtain social insects by tearing apart subterranean nests, mounds and decaying trees. Once the gallery is open, the echidna uses its sticky tongue to collect a maximum of individuals with each sweep. Echidnas do occasionally take insects on the surface, as indicated by the presence of remains of arboreal ants in echidna scats collected in Kakadu National Park. Presumably these ants were taken when on the ground.

Obtaining access to nests and galleries is not without cost. Various ant and termite species have evolved defences against predators. The ant genus *Calomyrmex*, for example, produces a secretion from its mandibular gland that has been shown to repel insectivorous vertebrates, such as the echidna. Similarly, termites of the family Rhinotermitidae use unpalatable defensive chemicals. Termites of the genus *Nasutitermes* are capable of squirting a

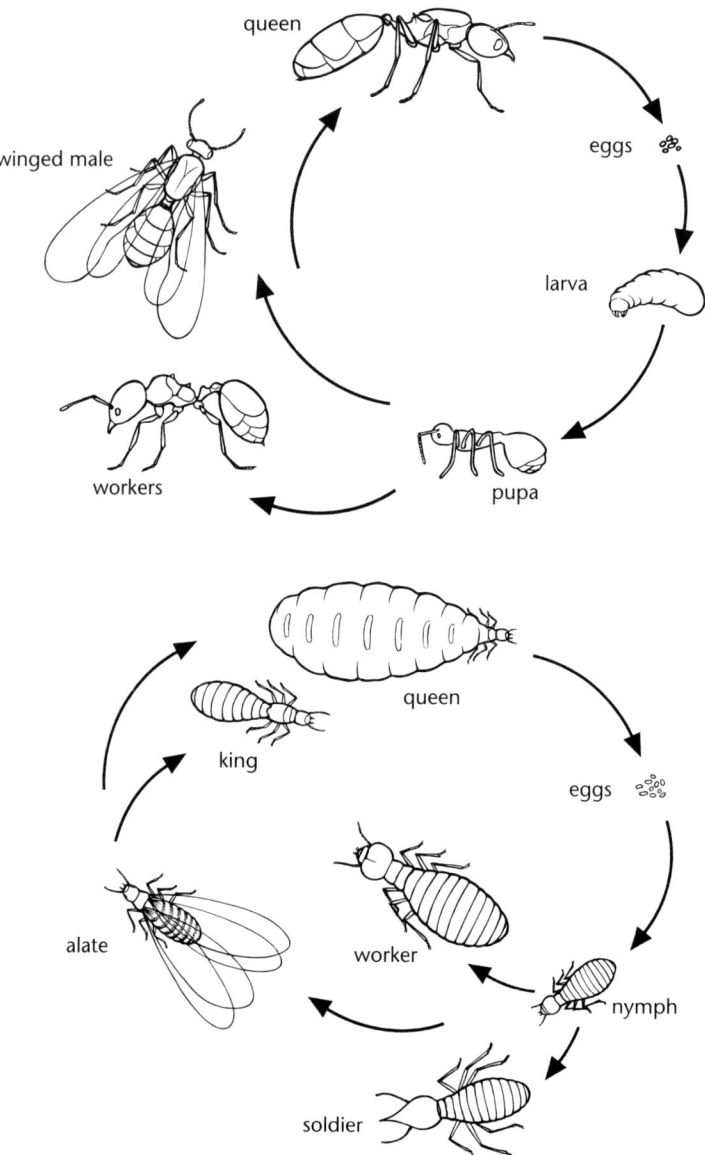

Figure 8.1 Life cycles of ants (top) and termites (bottom).

noxious secretion, although some species of this group are an important component of the echidna diet in some regions.

Ants that have stings, such as bull ants (*Myrmecia* spp.), and those with biting apparatus are also avoided. Echidnas avoid, if possible, the larger biting ants and go for the relatively defenceless prey – larvae, pupae, queen ants and winged males and females (Figure 8.1).

Meat ants, *Iridomyrmex detectus*, are particularly ferocious and attack in large numbers any intruder into their nests. These nests are in the form of low-domed mounds, roughly circular with a diameter up to two metres. The nests may also extend as much as two metres underground. During late winter and early spring, males and fat-laden virgin queens come into the upper galleries, congregating closer to the surface in sunny, northern aspects. Lipids account for 47 per cent of the dried weight of virgin queens, but only 18 per cent of workers and 10 per cent of males. At this time echidnas attack meat ant mounds, burrowing in with a delving action of the forelimbs and with hunched shoulders directed into the heart of the mound. From time to time they withdraw the head and then thrust it back into the hole again, putting up with the attack and bites of the workers, as clearly evidenced by the frequent and vigorous scratching of the head and chest region. Attacks on the meat ant mounds cease when the winged forms, males and queens, make their nuptial flight in October.

The impetus for echidnas to attack meat ant mounds in spring is probably a result of depleted fat reserves, especially after hibernation and for females who need to build up reserves for rearing young. In the 'off season' echidnas might take the odd meat ant worker on the surface but their nests are wisely ignored.

Food types and selection

Short-beaked echidnas eat mainly ants and termites; hence they are classified on the basis of their diet as myrmecophages (ant and termite specialists). Their diet also includes larvae of other invertebrates (especially scarab beetles), as well as adult beetles and earthworms. The size of the prey item is generally restricted by the gape of the mouth, which only opens about five mm – just enough to allow the tongue to dart out and return with small items stuck to it with a very sticky saliva.

Echidnas are opportunistic feeders, ingesting ants and termites as they are found with the probable exception of purposeful hunts for energy-rich meat ant queens and scarab larvae. It is possible to be both opportunistic and selective. Selection can come about due to habitat partitioning, hence cryptic subterranean and arboreal species are rarely encountered, or ignoring certain species when encountered, such as those with chemical defences. In Kakadu National Park, out of a probable 40 genera of ants present, 29 genera were found in echidna scats. As many as 14 genera were found in individual scats. In one area of Western Australia where 39 species of termites were present, echidnas were found to take all four of the species which occurred with high frequency, and most of the others except for a few rare species.

The balance between ants and termites in the diet reflects the balance in nature. Termites decrease as a proportion of the terrestrial and subterranean arthropod fauna from arid to mesic regions in Australia. Hence in cool, wet regions of Australia the echidna diet contains more ants than termites, while termites predominate in the diet in hot, dry regions.

All other factors being equal, echidnas prefer termites to ants. Termites are more digestible than ants, having a smaller component of their bodies as indigestible chitin exoskeleton. They also often live in larger colonies. However, termites of one entire termite family, Rhinotermitidae, which have well developed defence mechanisms and produce a variety of particularly noxious chemicals, are shunned by echidnas. Soldiers of many other termites produce less potent chemicals and have biting mandibles, and therefore echidnas seek out the succulent nymphs and queen termites. On the other hand, no termites have stings.

Scarab beetle larvae (*Sericesthis* sp.) can be an important, seasonal part of the echidna diet and are taken wherever they occur. These are far too large to be ingested whole and are crushed with a twisting movement of the snout and soft material is sucked up leaving some nutrient behind. A study in agricultural land in New England found scarab beetle larvae to be a major component of the echidna diet. In an analysis of 99 scats, the biomass intake was calculated to be 56 per cent ants, 37 per cent scarab beetle larvae and only 7 per cent termites. The study area may have had an atypically rich supply of scarab beetle larvae which preferentially feed on roots and organic material in pastureland.

Locating prey

An echidna observed in the wild appears to be wandering at random, searching every nook and cranny, especially along fallen logs that it comes across. It will often move its head from side to side making snuffling sounds, and intermittently poke its snout against or into the soil. This suggests that echidnas use their sense of smell to locate prey, but we really do not know how echidnas find their prey. We assume that they use a variety of sensory mechanisms, and they may also hear their prey moving in logs or underground by placing the snout on the log or ground. Using this latter method, sound is transmitted directly by bone conduction to the inner ear. Interestingly, the study of seasonal foraging activities of echidnas in the New England region found that energy-rich final instar scarab beetle larvae were only dug up by echidnas in the spring even though they were also present in winter and at least some echidnas were foraging in winter. The authors of that study suggested that this could be explained if echidnas located the larvae by sound because the larvae would not be active in winter.

Vision may also play a role, but an echidna can manage without the sense of sight because a blind echidna is known to have survived in the wild. In a field study in Western Australia two echidnas, each with only one eye, maintained normal body weight. Eyesight is of course relatively unimportant in locating subterranean prey. Eyes are at risk when digging and when exposed to attack by biting and stinging insects. Some protection against bites and chemical secretions of ants and termites and from continuous abrasion from dirt and sand is provided by keratinisation of the corneal epithelium. Of course echidnas might well close their eyes when attacking ant and termite colonies.

The echidna's snout possesses a variety of sensory nerve endings including those sensitive to touch, vibration and weak electric fields. No doubt they all play a part in making the snout a powerful and selective tool for detecting a variety of prey (see Chapter 5). The snout makes good electrical contact with the soil because its tip is always moist. The ability to detect weak electrical fields in soil probably helps echidnas find large items such as scarab beetle and cossid moth larvae to which they are particularly partial. Cossid moth larvae can be quite big – up to eight centimetres long – and as thick as a little finger. There is little doubt that an animal of this size produces detectable bio-electric activity as it moves through the soil.

Foraging activities

An echidna may simply trap ants by lying on top of an ant mound with its tongue extended on the surface of the mound. As the ants walk across its tongue, the echidna intermittently withdraws it into its mouth, along with the attached ants. This behaviour can continue for hours. An echidna walking through the scrub can use its forepaws and snout to turn over leaf litter on the ground and expose the insects underneath. The echidna also uses its powerful forepaws to tear apart rotting timber lying on the ground and expose termite nests (see photograph, page 28). Sometimes the echidna uses the snout in a 'cork-screw' action to turn over soft soil or pasture.

Foraging echidnas leave ample evidence of their activities – in the soil, near logs, in mounds, and in surface leaf litter:

- Semicircular excavations in the soil usually have a raised rim 4–20 cm, and often have a distinct echidna snout-shaped hole at the end. This

snout impression is commonly termed a 'snout prod'. Sometimes it is difficult to distinguish echidna diggings in soil from rabbit diggings, but usually they are wider as echidnas tend to push dirt out sideways more than other diggers. When concentrated in a small area, this activity can produce an area that has the appearance of ground rooted about by feral pigs.
- Often semicircular scoops with raised rims may be seen near logs. Echidnas commonly dig at the base of tree stumps and around hollow logs.
- When digging into ant or termite mounds echidnas can leave a simple conical hole up to 8 cm deep with a snout prod at the end of the cone, an echidna-sized chamber (sometimes with a snout prod at the end) penetrating to the upper brood chambers, or a tunnel with a rounded roof and flat floor up to a metre deep penetrating to lower brood chambers.
- Scratchings in the soil surface or in leaf litter are difficult to distinguish from those made by ground dwelling birds, such as lyrebirds. Echidna scratchings are relatively messy and usually are about 0.5 m across.

Strength

When foraging, echidnas often overturn fairly large rocks and rip apart large logs. One of the authors has observed that echidnas moved quite heavy pieces of slate paving, used as pathways in a domestic setting, in their search for ants. Stories of their ability to escape from captivity are legion (see photograph, page 101). This amazing ability was highlighted by the nineteenth century zoologist George Bennett who wrote that, 'when they (echidnas) are got, if the dissecting knife is not used at once the difficulty is to keep them'.

Their combination of persistence, burrowing ability and remarkable strength has led to echidnas being termed the 'escape artists of the animal world'. An echidna put into a box has been known to break through to the ground beneath it and tunnel three metres, coming out under a paling fence. Not only can echidnas climb trees, they can also scale two-metre-high cyclone fencing. Wire netting cannot hold them even if it is held in place by broad tacks. Their power allows them to move large stones weighing up to 13.5 kg. On one occasion an unwary zoologist left an echidna in his kitchen overnight and next morning found the refrigerator had been moved towards the centre of the room!

The echidna is a remarkably strong animal for its size. How does it generate such strength? Simply observing the animal reveals that it has short stout limbs. Anatomical dissection reveals well developed limb musculature attached

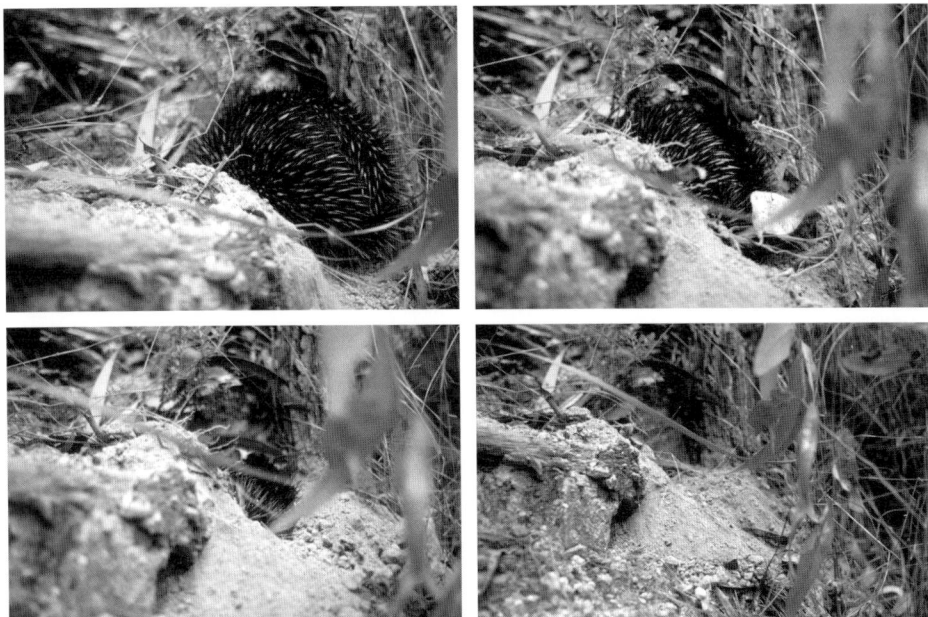

Going … going … gone! The echidna has a remarkable ability to disappear from view within seconds. Photo: Anne Musser.

to a solid shoulder girdle. The scapulas, clavicles, interclavicle, epicoracoid and coracoid bones form a solid yoke of bone which almost completely surrounds the chest and cervical vertebrae (see Figure 3.5). The buttressed shoulder girdle forms a very strong and stable base for the origin of the powerful forelimb muscles.

Figure 8.2 shows a comparison of mechanical advantage in the flexion of the elbow by the biceps in humans and echidnas. This mechanical arrangement is termed a third order lever. The fulcrum is the elbow and the lever is the combined radius and ulna bones. The biceps muscle has its origin at the shoulder on the coracoid and epicoracoid bones and its insertion largely on the shaft of the radius. Contraction of the biceps produces an effort on the radius and ulna which is transmitted to the forepaw where it exerts a load. The mechanical advantage is equal to the load divided by the effort. It can also be expressed as the length of the effort arm, that is, from the elbow to the insertion of the biceps, divided by length of the load arm, from the elbow to the forepaw. Thus the greater the length of the effort arm in relation to the load arm, the greater will be the mechanical advantage.

A comparison between humans and the echidna is quite instructive. In a human, the biceps inserts into the radius less than a quarter of the way along the shaft from the elbow, so the effort arm is quite short in relation to the load arm. In marked contrast in the echidna, the insertion of the biceps is almost

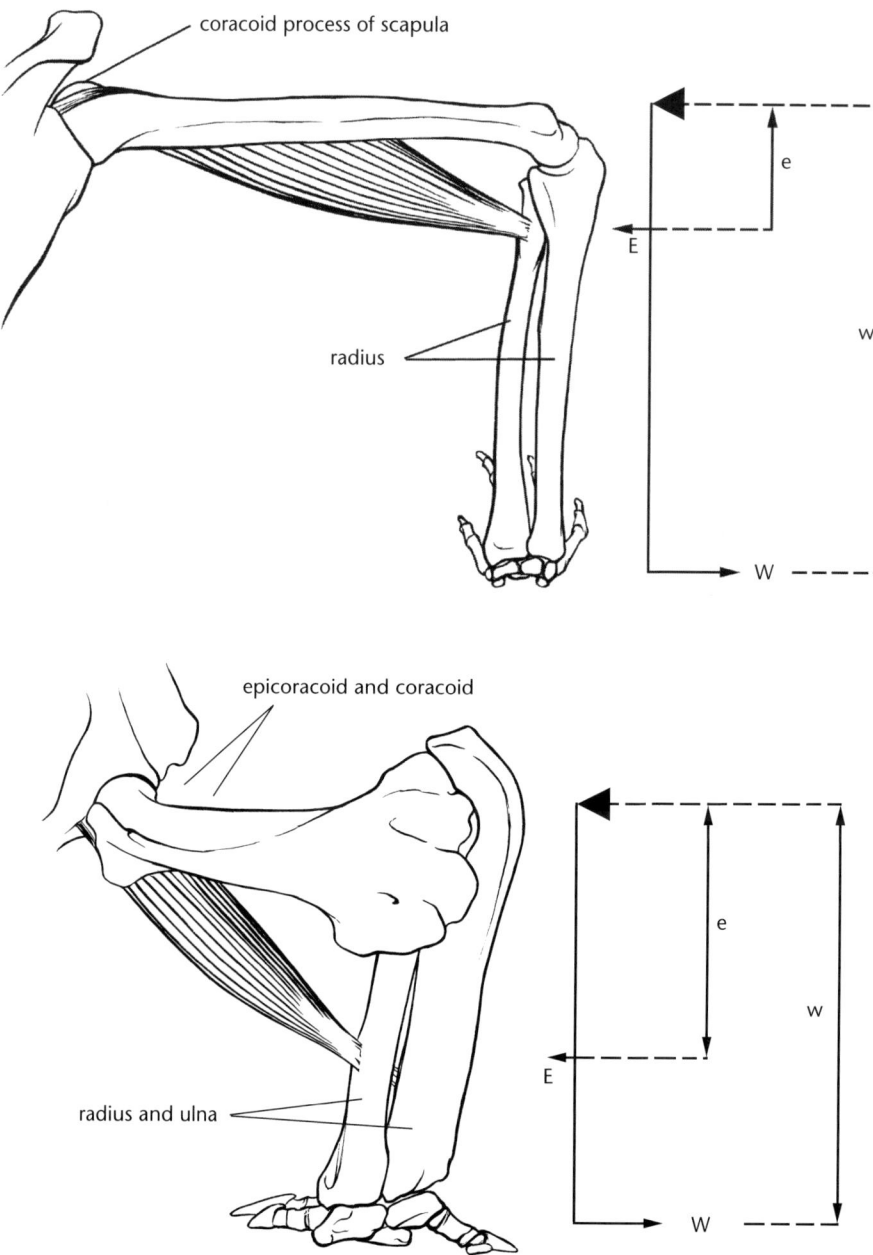

Figure 8.2 A comparison of mechanical advantage in the flexion of the elbow by the biceps in humans (top) and echidnas (below). This is a third order lever. The mechanical advantage (W/E) of the lever equals the length of the effort arm (e) divided by the length of the load arm (w). The longer the effort arm in relation to the load arm, the greater will be the mechanical advantage. Hence the greater mechanical advantage of the echidna. ◄ indicates the fulcrum.

halfway along the radius from the elbow which results in a greater mechanical advantage in the echidna compared to the human being. In the echidna limbs, the muscle mass is large in relation to the short bones which also results in more powerful function. The squat humerus possesses broad areas for attachment of the thick muscles. The echidna's forepaw possesses five spatulate claws which break up the soil and, along with the palm, provide a very efficient digging tool.

The snout

The snout plays a critical role not only in the detection, but also in the retrieval of prey, as in the turning over of softer soils. To fulfil this function the snout must be able to penetrate the soil repeatedly and efficiently. A closer examination of the snout reveals that it is shaped like a double wedge (Figure 8.3). The tip of the snout forms a relatively blunt wedge but the main length of the snout only widens gradually from nostrils to the base adjacent to the eyes. The mechanical advantage of a wedge equals the load force applied bilaterally by the wedge against an object divided by the effort force driving the wedge into the object. Mathematically it can be shown that the mechanical advantage is equal to the slant height of the wedge divided by its thickness. Thus the more

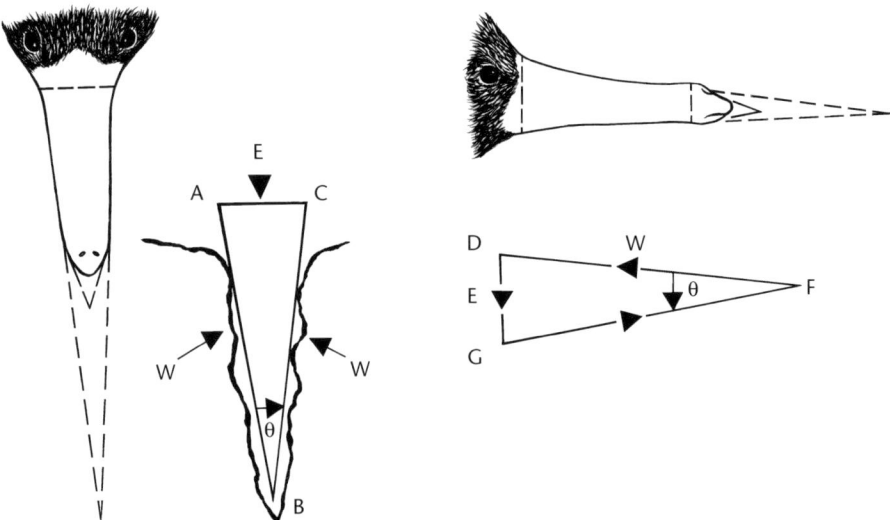

Figure 8.3 The snout as a wedge and its mechanical advantage. The mechanical advantage of a wedge equals the slant height of the wedge divided by its thickness. Thus the more acute the angle (θ) of the wedge, the greater its mechanical advantage. The approximate mechanical advantage of the tip of the echidna snout is 1.5 and of the main length of the snout is 6.0. The effort (E) is multiplied to a 6X greater lateral force (W) to split open the soil.

acute the angle of the wedge, the greater will be its mechanical advantage. Initially, when the first centimetre or so of the snout enters the soil, the mechanical advantage is quite small, but once the main length of the snout enters the soil, the effort is multiplied six times. The main length of the snout then tends to split the soil apart and in firmer soils this splitting will occur ahead of the advancing blunter tip of the snout, protecting it from major trauma.

The gape of the echidna's mouth is only about 5 mm. Some worms and larvae are too big to pass through the mouth whole. The echidna uses its snout to crush the prey at the bottom of the hole it makes while searching for its prey.

The tongue

The echidna's tongue is a remarkable organ (Figure 8.4). It is oval-shaped in cross-section and about 15–18 cm long when fully extended. It can be shot out of the mouth with lightning-like speed, to 100 times per minute. This is how the echidna got its name of *Tachyglossus* or 'fast tongue'.

The tongue is in two parts – an extensible anterior portion and a prominent, fixed posterior portion. The anterior part is circular in cross-section and is extremely agile. It is particularly flexible at its tip, which can be bent into a U-shape, which allows it to track actively inside the narrow galleries of ant

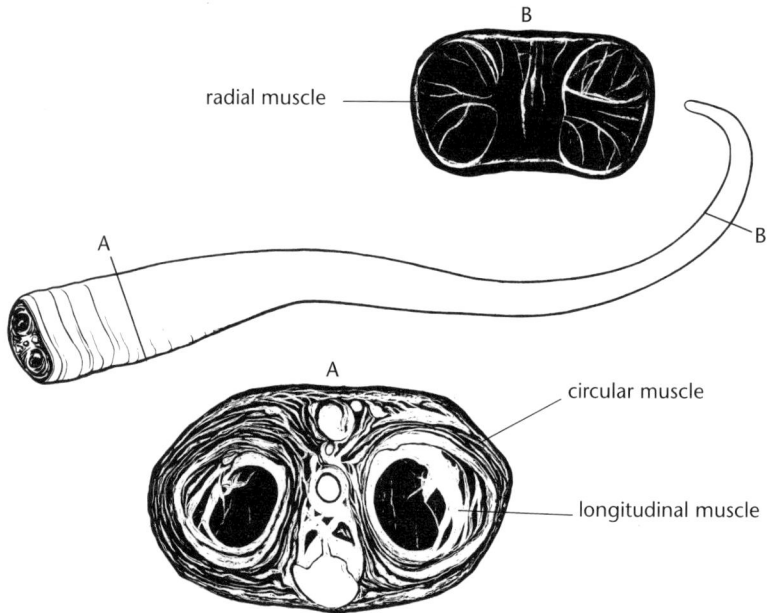

Figure 8.4 Anatomy of echidna tongue musculature. The extended echidna tongue is 15–18 centimetres long. The points at which the two cross-sections have been taken are indicated by straight lines across the tongue.

and termite nests and to catch scurrying individuals. Why an echidna does not end up with a tongue full of splinters when feeding from termite nests in broken logs is unknown.

How does the echidna extend its tongue? Towards the back of the tongue there are two bundles of longitudinal muscles side by side, which are surrounded by circular muscle. In order to extend the tongue rapidly, blood spaces in the tongue are engorged and the circular muscle contracts at intervals along its length which results in the extension or forward 'squirting' of the longitudinal muscle. Retraction is produced predominantly by relaxation of the circular muscle and active contraction of the longitudinal muscle. However, some of the energy for retraction of the tongue comes from the conversion of potential elastic energy due to the deformation of the tongue's shape back into kinetic energy as the tongue resumes its resting position. The remarkable agility of the tongue's tip is due to muscle fibres that are arranged in a radial fashion. These radial fibres split up the longitudinal muscle into six to eight smaller muscle bundles.

At the very back of the tongue are the echidna's taste buds. These are mammalian in structure with circumvallate papillae.

Uptake and mastication of prey

The way in which echidnas open the mouth is unusual. Instead of the hinge-like action of the jaws of most mammals, the thin mandibles (left and right halves of the lower jaw) rotate about their long axes and swing outward enough to stretch open the mouth.

When the echidna extends its tongue, the intact or crushed prey adheres to its surface. The stickiness of the tongue is due to it being coated with a secretion from the sublingual salivary glands. The sublingual glands empty their secretion – which has the consistency of treacle – by numerous ducts which open along the greater part of the floor of the mouth. Thus the echidna has an efficient mechanism that is provided for coating the long tongue with the sticky secretion.

In the absence of teeth, the job of mastication in echidnas has been taken over by 'soft' tissue substitutions comprised of two sets of hardened, keratinous spines (Figure 8.5). One is a transverse set on the roof of the mouth (the palatal series). The other is a matching set on a raised knob (the dental pad) at the base of the tongue (the lingual series). These occlude to grind the hard insect exoskeletons into a digestible paste. When the echidna retracts its tongue, the prey are transferred to the back of its mouth and are scraped off onto the dental pad by the back-and-forth action of the opposing palatal and lingual spines.

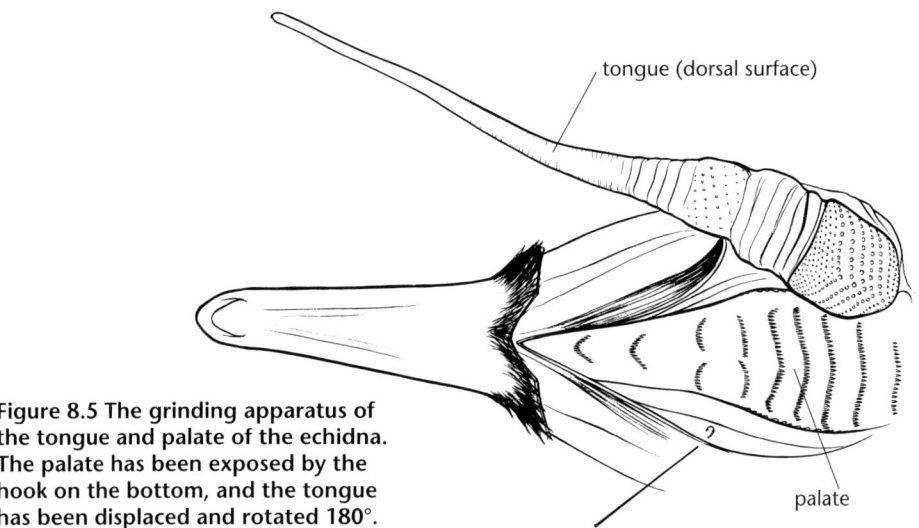

Figure 8.5 The grinding apparatus of the tongue and palate of the echidna. The palate has been exposed by the hook on the bottom, and the tongue has been displaced and rotated 180°.

This feeding process is very efficient. A fair sized echidna of about three kilograms can ingest 200 g wet weight of termites in 10 minutes.

Digestion

The fragmented food then passes through the oesophagus to the stomach. The lining of the echidna's stomach is most unusual. It contains no secretory glands but consists of cornified stratified epithelium similar to horny skin. The pH of the echidna stomach is not low (acidic) as it is in most mammals, but is close to neutrality (6.2–7.4). The gastric peristalsis of this rugged stomach lining grinds dirt particles and fragments of insect together, producing a very effective mechanism for the further breakdown of hard insects. The stomach of an echidna is quite elastic and can hold an enormous number of ingested insects. There are Brunner's glands (found only in mammals) in the pyloric region.

Digestion takes place in the small intestine which is approximately 3.4 m long. The pancreatic and bile ducts are combined into a single duct which enters at the start of the small intestine. The passage of the bowel contents is slow, allowing thorough digestion of the soft parts of the prey. The broken pieces of insect exoskeleton are not digested but pass out in the faeces or 'scats'.

Scats provide an invaluable record for zoologists wanting to know what an echidna eats. As well as fragments of exoskeleton, the scats contain soil and sand which is ingested with the prey. Echidna scats have a characteristic cylindrical shape (see photograph, page 29) and are quite firm once the water content has evaporated. Scats are not common even in areas where echidnas are known to abound. Echidnas in captivity have been seen to cover their scats. They may do this in their natural habitat as well, although echidna scats are sometimes seen on the surface.

9
Metabolism

Some time ago mammals were known as 'warm-blooded' animals as opposed to 'cold-blooded' animals such as lizards and sea urchins. Live echidnas have been found in the Snowy Mountains in winter with body temperatures below 5°C, which illustrates why the term 'warm-blooded' has fallen out of use. However, echidnas have also been found walking through the snow in the cold with body temperatures near 32°C, so there are clearly some major differences between echidnas and lizards and sea urchins.

Torpor and hibernation

Torpor and hibernation are closely linked phenomena in endothermic animals – those that can maintain their body temperature above that of the environment. Torpor is fundamentally a lessened ability to make locomotor responses. It can be a response to low temperature when there is a short, often daily, abandonment of normal body temperature range of the animal (euthermia) in which it becomes inactive and its body temperature, heart rate and metabolism fall to levels below those seen either when the animal is active or when it is simply asleep. Unlike awakening from sleep – which is simply a matter of turning the nervous system on to full consciousness and getting up and away in an instant – arousal from torpor is much slower, requiring a relatively gradual return of metabolic and circulatory functions to their normal level.

Long-term torpor is a feature of hibernation. An animal in hibernation is in a deep torpor in which body temperature can approach the ambient temperature. However, hibernation refers to a much broader phenomenon and includes brief periods of arousal between repeated bouts of torpor. Hibernation has traditionally been associated with winter. Similar bouts of torpor in summer are termed 'aestivation'.

Echidnas show bouts of shallow torpor at any time of the year as well as a pattern of back-to-back bouts of torpor similar to hibernation.

What is often termed 'classic hibernation' is taken from a massive literature on studies with rodents, bats and a few other species of animals living in the northern temperate and arctic regions. In classic hibernation, body temperature drops to within a couple of degrees of ambient temperature, heart rate slows dramatically and metabolic rate is at absolute minimum. This clearly conserves energy in the face of cold winters when food is unavailable. An apparent contradiction to the energy-saving role of hibernation is the fact that all classic hibernators have periodic arousals in which they return briefly to normal conditions of body temperature, heart rate and metabolic rate. Echidnas fit the physiology of classic hibernation (Table 9.1). In the Snowy Mountains, the echidna's body temperature has been found to drop as low as 4°C. Heart rate drops to 4–7 beats per minute and respiration drops to 0.3

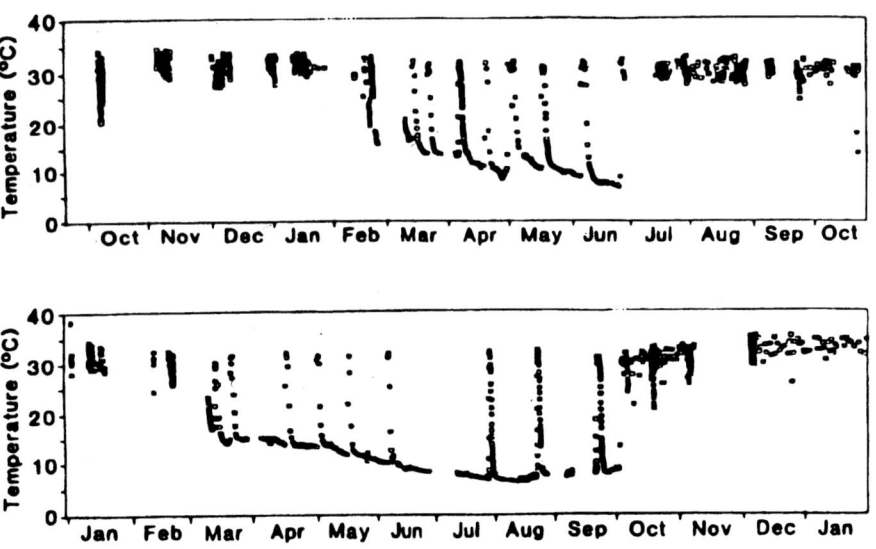

Figure 9.1 Body temperatures of two echidnas in hibernation recorded by radio-telemetry in the Snowy Mountains. Note the distinct hibernation season in each case (from Grigg et al. 1992).

breaths per minute. Metabolic rate can be less than 12 per cent of euthermic levels. Like classic hibernators, echidnas exhibit periodic arousals from hibernation, during which time they may actually move to another site (Figure 9.1).

Table 9.1. The pattern or hibernation in echidnas. These values are taken from studies carried out in different parts of Australia. Considerable variation may occur depending on environmental conditions.

Entry (February–April)
    Preceded by reduced food intake even though food is available
    Preceded by test drops (brief torpors with decreasing drops in body temperature) Mar–May
    Males enter hibernation first
    Reproductive females enter hibernation later than males

Hibernation
    Minimal physiological values:
        Body temperature: 4°C
        Heart rate: 4 beats per minute
        Oxygen consumption: 0.02 ml O_2 per gram body weight per hour
        Nitrogen excretion: 40–50 mg per kilogram body weight per day
        Weight loss: 2.07 grams per kilogram body weight per day
    Number of torpor bouts: 13 (highly variable)
    Duration of individual torpor bouts: mean of about 11 days
        Torpor bouts become progressively longer at the start of hibernation
        Arousals become more frequent towards the end of hibernation
    Duration of periodic arousals: 1.2 days
        Echidnas occasionally move to another site during arousals
        Arousals correlate with rises in temperature during brief warm spells and echidnas in the same area tend to arouse at the same time
    Time taken to achieve euthermia: 8.8 hours

Termination of hibernation
    Males terminate first – mid June
    Reproductive females terminate July/August
        Mate within two days of ending hibernation
    Immatures and non-reproductive females may hibernate until September/October

Euthermia
    Incubating and nursing females are nearly homeothermic when in the nursery burrow
    Daily body temperature variations of 4°C are common
    Echidnas may have torpor bouts at any time of the year
    Weight gains of up to 18% can occur out of breeding season

Hibernation takes place in all climatic zones – it has been observed in the Snowy Mountains, New England tablelands, Tasmania, south-east Queensland, Mudgee (New South Wales), and Kangaroo Island. The Snowy Mountains represent the coldest winter habitat and Kangaroo Island is probably the mildest with winter minimum temperatures well above freezing.

When it comes to the timing of hibernation, echidnas do not fit the neat pattern of classic hibernation where torpor correlates with avoidance of low ambient temperatures and food scarcity. The confounding features of hibernation in echidnas are:

- males may enter hibernation in February – it is still summer and food is abundant,

- males come out of hibernation in the middle of winter,
- food is available all year around except, perhaps, above the snowline,
- mating is in winter,
- reproductive females come out to mate in late winter,
- immatures and non-reproductive females might sleep on until October (Spring), and
- hibernation occurs in regions with mild winters.

Because echidnas are clearly not using hibernation entirely to avoid cold or times of food scarcity, the interesting question arises: why do echidnas hibernate?

There are four different approaches to the answer in the literature. One proposes it is part of an optimal foraging strategy, possibly an adaptation to the energy poor and climatically unpredictable Australian continent; another proposes that it is part of an overall low energy strategy; a more traditional proposal, going back to the early 1900s, is that torpor, and hence hibernation, in echidnas is plesiomorphic, a reflection of an ectothermic ancestry; and finally we propose that the overriding factors in the timing of hibernation in echidnas are the imperatives of the reproductive cycle.

In our opinion the question is not 'Why do echidnas abandon endothermy and enter a state that is fundamentally ectothermic?', but rather the question should be 'Why do echidnas sacrifice the advantages of ectothermy, living in the slow lane, for endothermy?'. Taking the long, evolutionary view, and assuming that mammals evolved from ectothermic ancestors, we suggest the best answer is that endothermy offers great advantages for reproduction. After all, there are many examples in the animal kingdom where complex and expensive structures (antlers of elk, tail feathers of peacocks) and functions (long courtship, milk production) have arisen because they have a direct, positive adaptive value for reproductive success. Selection has been shown by many authors to have its greatest outcome when resulting in direct positive effects on reproduction. The reproductive benefits of endothermy are probably widespread, but the most obvious are the benefits for development and raising of offspring. Higher incubation temperatures speed up development. Higher activity potential for parents facilitates the feeding of offspring and provision of parental care.

In the annual cycle of echidnas, endothermy is taken up when required for mating and raising of offspring, otherwise much of the annual cycle is taken up by torpor bouts. Hence males, having made their only contribution at mating, go into torpor relatively soon thereafter. They wake up earlier than females presumably because it is necessary to complete spermatogenesis before mating. Immature echidnas and non-reproductive females continue hibernation into September or October. Non-reproductive females can spend seven months of the year in hibernation. Reproductive females need to arouse and mate in mid-winter because of the time required in this species for gestation, incubation of the egg, and growth of the young to a size and developmental stage that allows survival though the next winter.

Sites chosen as hibernacula always provided 100 per cent cover (Table 9.2). In the Stanthorpe region of southern Queensland rabbit burrows were most commonly used as hibernacula. In the part of the Snowy Mountains where the studies on hibernation were made, rabbits are relatively scarce and only one radio-tracked echidna hibernated in a rabbit burrow. In Tasmania most echidnas simply dug themselves into the soil.

Table 9.2. Sites used as hibernacula in the Snowy Mountains of New South Wales. Numbers refer to the number of times a site of that type was observed to be used.

Cavity in base of tree stump (6)
Hollow log (4)
Wombat warren (4)
Natural underground cavities (2)
Cavities formed by living tree roots (2)
Cavity in man-made dirt or rock piles (2)
Rabbit warren (1)
Soft centre of partly dead tree (1)
Soft centre of partly dead stump (1)
Soft earth under fallen tree (1)
Pile of dead trees (1)
Debris at base of tree (1)
Base of termite mound (1)

Overall, the remarkable feature of thermoregulation in echidnas is not that they enter torpor and hibernation but that they are active and mating at the coldest time of the year and that they are capable of numerous endothermic arousals during the hibernation period. These arousals are achieved without brown fat, which is present in placental hibernators but absent in monotremes and marsupials, and represent a remarkable burst of metabolic activity for an animal which C.J. Martin, writing in 1902, considered to be so inferior to other mammals that it 'abandoned all attempts at keeping warm in winter and sank into a reptilian state'.

The above details are of course for short-beaked echidnas. Recently Gordon Grigg and his colleagues studied two long-beaked echidnas held at Taronga Zoo in Sydney. They were unable to find any evidence of long-term torpor, while two short-beaked echidnas in the same pen underwent hibernation in the cooler months and short-term torpor throughout the year.

Body temperature and metabolic rate

Echidnas were long assumed to have metabolic capabilities inferior to other mammals for two reasons. First, they share with the platypus the distinction of having the lowest body temperature of any mammal. Even when active, an echidna's body temperature is about 32°C and normally never goes above 34°C (Figure 9.2). Laboratory studies have shown that echidna body temperatures are highly variable, depending on the activity level and the temperature of their surroundings (Figure 9.3), and under laboratory conditions echidnas seem to depend entirely upon heat production by contraction of voluntary (striated) muscle to maintain their body temperature above 30°C. If inactivity continues for several days, body temperatures can fall to near 20°C. Free echidnas can seek insulated microclimates, such as burrows, in a complex temperature environment and can use behavioral means, such as basking, to thermoregulate. Radio-tracking studies in the Snowy Mountains did not show the close relationship between air temperature and body temperature shown in the laboratory, except under extreme conditions.

Data from field studies in the Snowy Mountains carried out by Gordon Grigg and Lyn Beard show that female echidnas can be homeothermic endotherms in the true sense, regulating their body temperature tightly while they are incubating their eggs.

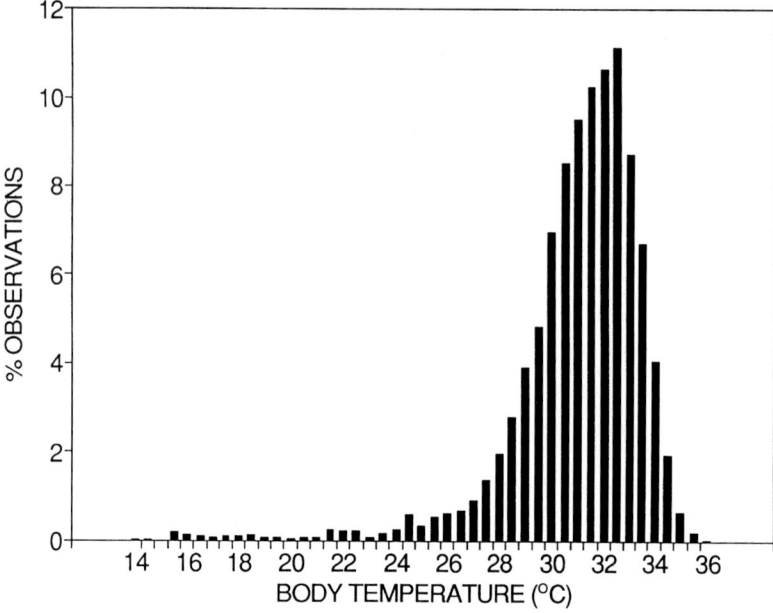

Figure 9.2 Frequency distribution of all body temperatures measured by radio-telemetry during the active season for echidnas in the Snowy Mountains (from Grigg *et al.* 1992).

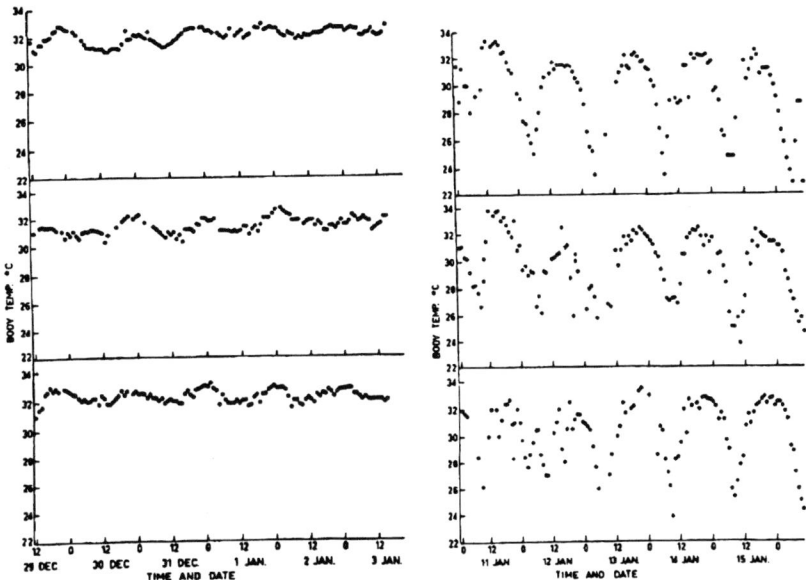

Figure 9.3 Body temperatures measured by radio-telemetry in laboratory studies of three echidnas under controlled conditions. The animals were maintained at a constant temperature of 25°C (left) and then 10°C (right). Variations in body temperature correlate with activity patterns. Body temperatures start to decline when the animals become inactive and start to rise after they become active.

Second, echidnas have a metabolic rate about 30 per cent of that of placental mammals, which, when torpor bouts are taken into consideration, gives them the lowest annual energy turnover of any mammal. Platypuses also have a lower metabolic rate than placentals, equal to about 50 per cent. The usual way in which such comparisons are made is to determine the amount of oxygen consumed by an animal just to stay alive when at rest. The rate at which oxygen is used is then compared with other animals per unit weight (it is most unfair to compare one echidna with one mouse or one elephant). On that basis it takes only one-third as much oxygen to keep one gram of echidna ticking over as is required for one gram of dog, cow or human. Marsupials fall somewhere in between.

It is usually assumed that this difference in metabolic rates reflects an evolutionary gradient. However, other factors may be involved. B.K. McNab studied the metabolic rate of a variety of ant- and termite-eating mammals, including a few marsupials and three short-beaked echidnas (two from Tasmania and one probably from New Guinea). He concluded that there is considerable physiological convergence among these species and that adaptations related to feeding could account for most of the lowered metabolic rate that is typical of ant- and termite-eaters. McNab also examined a wide variety of burrowing mammals and also concluded that there was a tendency towards reduced metabolism in this group but only in species with body weights over one kilogram. The reduction related to burrowing was not as great as that related to ant- and termite-eating, but echidnas of course qualify on both counts. However, it is difficult to explain the fact that platypuses and long-beaked echidnas (the latter earthworm-eaters and non-burrowers) also have very low metabolic rates. Low metabolic rate, and consequently low body temperature, seems to be a characteristic of monotremes, and it is highly unlikely that the lifestyle of the echidna is the sole basis for its low metabolic rate. Nor is a low metabolic rate necessarily a disadvantage; after all, it takes roughly two-thirds less energy and therefore less food to keep an echidna going than is required for a similar sized placental. Echidnas achieve further savings in winter in cold environments by turning the metabolic furnace down, not up.

Temperature regulation

In most of Australia and New Guinea, where echidnas live, they rarely face cold. In many parts of their geographical range environmental temperatures below 15°C are rare. The problem faced by echidnas in such places is far more likely to be heat, and echidnas have little tolerance for heat. Body temperatures above 34°C are probably lethal, yet echidnas have no sweat glands and do not pant. The ability of echidnas to maintain an active body temperature of about 32°C depends in part upon circulatory adjustment. Blood can be directed towards or away from the skin to facilitate or restrict heat loss. But the main defence echidnas have against heat is avoidance. In hot climates they seek cool microenvironments such as rock caves, wombat or rabbit burrows, and hollow logs. They restrict their activities to night or the morning and evening hours. Echidnas have also been observed to swim in dams and creeks in summer, although they also live in many places where water is absent.

In animals that are able to actively control their body temperature independently of ambient temperature (birds and mammals), there is an ambient temperature at which it only takes minimal effort to maintain normal body temperature. This is known as the 'thermoneutral zone', much the same as the 'comfort zone' for humans. The echidna thermoneutral zone is around 25°C, at which point metabolic rate as measured by oxygen consumption is minimal.

Water balance

Although echidnas lack sweat glands and do not pant, they lose water with every breath as water vapour passes through the respiratory system. The snout is probably very important in limiting this water loss by a mechanism known as 'countercurrent heat exchange'. As air is inhaled through the snout it passes through a complex bony labyrinth which provides contact with a large surface area of moist tissue (see Figure 5.6). The dry incoming air evaporates water from this tissue, cooling the walls, and becomes warmer and moister as it passes deeper into the respiratory system. On exhalation, this moist air moves back along the passages in the snout whose walls are progressively cooler towards the outside. As air cools it becomes saturated with water vapour until water is deposited on the passage walls near the snout. On the next breath, this water is used to saturate the incoming air and again cools the walls, continuing the water conservation cycle.

Another major route of water loss is by way of the urine. The kidney of mammals serves to minimise this loss by concentrating waste products. Echidnas have a basically mammalian type of kidney, a tubular-loop kidney, and achieve the same level of concentration as rabbits and dogs, although echidnas do not attain the high levels of urine concentration and hence water conservation as desert rodents. They do not have kidney tubules with elongated loops (the so-called Loops of Henle) that are found in the kidneys of desert rodents and which allow resorption of water. Echidnas do have, scattered among mammalian nephrons, dwarf nephrons typical of lizards and snakes.

A small amount of water is also lost by evaporation across the skin. Some, too, is lost in faeces, although echidnas in the wild usually produce relatively dry droppings.

For a three kilogram echidna maintained in a laboratory in dry air at 25°C, water loss is calculated to be:

- 51 grams per day from the respiratory tract and skin,
- 60 grams per day in urine, and
- 9 grams per day in faeces,
  giving a total loss of 120 grams per day.

The important question arising from such studies is whether echidnas can make up that water loss without drinking. This can be calculated by considering the diet. Termites are about 77 per cent water, and one laboratory study found that an echidna would eat about 147 g of termites a day. That intake would provide 113 g of free water. The breakdown of fat and protein molecules from these termites would provide an additional nine grams of water, giving a total water input of 122 g. It is close, but the water balance under these conditions is positive and the 120 g of water lost can be replaced. The balance would be tipped by increased activity, higher environmental temperatures or any change in diet that decreased water intake.

Echidnas have been observed to drink from open water in the wild. However, many echidnas live where open water is unavailable for at least part of the year. They may use a trick that has been well described for desert lizards: licking the early morning dew from moist surfaces. Echidnas have been observed to lick droplets of water that have accumulated on plants as a result of dew, rain or fog. Whatever they do to tip the balance of water conservation towards survival, their continued existence in arid zones depends on selection of suitable microhabitats and avoidance of heat as much for the purpose of water conservation as for temperature regulation.

The principle nitrogenous constituent of echidna urine is urea, which is typical of mammals. The urine and the blood contain very little uric acid.

Circulation and respiration

Heart rate of an echidna at rest within its thermoneutral zone is 50–68 beats per minute, rising to 135–145 beats per minute when very active. The echidna's heart has four chambers and its red blood cells have no nucleus and are biconcave, as in all other mammals (and only mammals). Respiratory rate at rest is 5–6 breaths per minute. The lungs are large, with two lobes on the left and one on the right. The diaphragm is complete and muscular at the edge with a tendinous centre.

Echidnas are very good burrowers. They can dig more than a metre into meat-ant mounds during which time they must be subjected to a considerable degree of asphyxia – an interference with breathing resulting in a combination of low oxygen and high carbon dioxide concentrations in the blood. Echidnas have also been found buried up to 25 cm below the surface. By digging into the ground they are known to survive bush fires passing overhead. Echidnas have been observed to swim and we have seen one dive voluntarily underwater when approached by a human. It is not surprising then to find that echidnas are remarkably tolerant to low oxygen and high carbon dioxide levels. Laboratory studies have found that the echidna exhibits heart and blood vessel responses during simulated diving similar to marine mammals such as the seal. The heart rate slows down markedly to about 12 beats per minute and the peripheral circulation is also reduced profoundly. Similar but more modest changes occur in echidnas under soil. These adjustments are believed to conserve oxygen for the organs that are most sensitive to lack of oxygen, namely the heart and brain. Thus the echidna is well equipped to survive acute asphyxia which might occur in emergency situations such as burrow collapse, bush fire or flash flood.

Endocrines

There has been a consistent pattern for research on monotreme function to be triggered by anatomical studies that have shown organs to be different from those species considered to be true mammals such as the much-studied laboratory rat. Research into adrenal gland function is typical. Histological studies early in the twentieth century indicated that the overall arrangement of the adrenal gland was 'reptilian' in nature. There followed studies in the 1960s and 1970s that showed that the echidna's adrenal gland secreted the usual mammalian hormones (corticosterone and cortisol), was under control of ACTH from the pituitary and in general responded to stress as does the adrenal in eutherian mammals. Secretion rates are relatively low and the effects of removal of the adrenal glands minimal, consistent with the low metabolic rates and life in the 'slow lane' typical of echidnas.

Thyroid gland function was studied for the same reasons but with added interest because of the fundamental role of thyroid hormones in classical hibernation. The structure of the gland and the products produced are typically mammalian, but echidnas have low levels of thyroid hormones compared to placentals, although the platypus has high levels. Levels of thyroid hormone are depressed during hibernation which is odd because the thyroid gland has been reported to be enlarged in hibernating echidnas.

Other endocrine glands have been well described anatomically but have been subjected to very little functional study. In general the pancreas, pituitary, parathyroids and thymus are basically mammalian with relatively minor 'reptilian' characteristics.

10
Conservation and management

Conservation status

In New Guinea, both the long-beaked echidna and the short-beaked echidna are threatened by the twin disasters of increasing human population and imported Western technology. Old taboos, which in some cases precluded the killing of echidnas, began to break down with the spread of missionary influence. Both *Tachyglossus aculeatus lawesii* and *Zaglossus bruijni* probably had a patchy distribution in New Guinea before the arrival of Europeans. Now they are known to be common in only a few localised areas and there are many areas where New Guineans claim they were once hunted but are now gone. They are hunted for their meat which is highly prized for its oiliness.

Killing an echidna is not difficult because the defence responses of an echidna, such as partially burying itself in the soil, are totally ineffective against human hunters. The problem for the hunter, as for the field biologist, has always been to find the echidna. However, this is overcome by hunters using dogs. Dogs are extremely good at sniffing out echidnas and only a minimum amount of training is required to make this ability work for the hunter.

In recent years visiting zoologists have been unable to find echidnas in the highlands of New Guinea, although the occasional carcass has been spotted in native markets. The situation is critical for *Zaglossus bruijni* because the distribution of the species is limited to New Guinea and therefore the species must be considered to be in great danger of extinction. Being an earthworm eater and not an anteater, *Z. bruijni* is highly vulnerable to clearing of forest. It is protected by the governments of Papua New Guinea and Indonesia but hunting by traditional methods is allowed.

On mainland Australia, and especially on Tasmania and Kangaroo Island, short-beaked echidnas are common. Fortunately, they are spread through all major habitats, from coastal forests to central arid regions. They have managed to hang on in areas cleared for agricultural purposes as long as some islands of bush have been left untouched. Echidnas are still reported in the outer suburbs of all major Australian cities.

In the long term, habitat destruction and degradation will continue to lead to local extinctions. This is most likely to occur in areas where all native cover is removed to allow the mechanised processing of a single crop such as wheat. In forested areas the importance of logs and dead trees to echidnas (as well as many other native animals) is often overlooked in management of timber cropping and in application of excessive, frequent burning regimes. In the case of land cleared for grazing stock, felled timber left to lie on the ground provides a source of food and shelter for echidnas. If, however, the logs are simply pushed into a pile and burned, the habitat is destroyed.

Overall, the species has the security of being well established in a wide variety of habitats. Relatives of the echidna might have survived the events that lead to the extinction of the dinosaurs and they may well survive the effects of humans.

Predation

Like other terrestrial vertebrates that are relatively large and well protected (by spines in the case of echidnas), predation operates almost entirely against the young. Natural predators are mainly reptiles: snakes and large lizards. Goannas are good diggers with an efficient sense of smell and are therefore likely to have been the main killers of young echidnas until the late arrival of dingoes and relatively recent introduction of cats, dogs, foxes and pigs. We have been aware for many years of stories of dingoes killing adult echidnas by rolling them onto their backs, and echidna spines have been reported from gut contents of feral pigs as well as dingoes and foxes, but it is impossible to determine if this is a result of active predation or scavenging from carcasses such as road kills. It seems unlikely such activities could represent a major problem for echidnas. The only quantitative data on predation comes from the long-term studies on echidnas on Kangaroo Island (*T.a. multiaculeatus*) carried out by Peggy Rismiller. She found that of 22 echidnas being tracked before weaning, three were killed by goannas and four by cats. Of 17 sub-adult echidnas that had moved into her study area from outside, two were killed by cats. It is important to note that Kangaroo Island is free of foxes and it is likely, although unproven, that on the mainland foxes are equal to cats as predators on young echidnas. Certainly Tasmania, which has been free of foxes, appears to have a much greater population density of echidnas than comparable mainland areas.

Humans in automobiles are a threat in some areas, although in surveys of road-kills echidnas are never common. This is probably due to driver concern for tyres rather than concern for wildlife. On the other hand, the majority of injured echidnas handed in to wildlife rescue organisations have been hit by cars. A study of the cause of death of 73 echidnas from 1976 to 1985 in Victoria showed that 24 were caused by automobiles and 11 possibly by attack from dogs or foxes. One provision that can be taken to reduce road kills, especially in national parks, is provision of culverts under roadways. During field studies in Kosciuszko National Park echidnas were observed to make frequent use of drainage culverts under the busy Snowy Mountains highway.

Disease

Various handbooks and references for zookeepers provide alarming lists of diseases and parasites of echidnas. However, it is difficult for the general biologist to determine which of these are due to maintenance in captivity and which are apt to occur in natural populations. In a review in 1992, Richard Whittington considered this problem and proposed that the following diseases are of potential significance to wild echidna populations.

Flatworms

Spirometra erinacei, a flatworm (cestode) that parasitises the intestines of dogs (including dingoes), cats and foxes in Australia, has been found in several wild echidnas. This parasite is particularly dangerous, probably fatal, to echidnas because they are not adapted to it. *S. erinacei* almost certainly arrived in Australia with introduced carnivores. The link to the echidna is by the way of small fresh water organisms called copepods. Copepods carry an intermediate stage which infects them through faecal contamination of the water by an infected carnivore. Echidnas take in the infected copepods while drinking water. Feral cats and dogs as well as dingoes and foxes are found in most parts of Australia where echidnas are found, although the potential for infection is probably greatest in the semi-arid zones where a limited number of waterholes attract many animals.

Viral infections

Systemic viral infections similar to herpes have been reported in echidnas. Herpes viruses are important pathogens in many mammal species, but considerably more research is needed to determine if there is a host-specific echidna herpes virus or whether cross-contamination from other animals occurs. Pox virus has been detected in echidna skin and may be the agent responsible for flaky skin (proliferative dermatitis) that is a frequent problem with echidnas in captivity.

Protozoal infections

Protozoal infections are commonly detected on autopsy of both captive and wild echidnas. *Eimeria* and *Octospotella* may reach high levels in intestinal epithelium but do not usually cause signs of disease. Other protozoa can infect soft tissues such as lungs, where they cause pneumonia. As with viruses, work on identifying specific protozoa that infect echidnas is only beginning.

Other parasites

Whittington listed eight species of fleas for echidnas, including five species of the genus *Echidnophaga*. One mite is recorded, *Odontocarus echidnus*, and 10 species of ticks are listed, one of which is *Amblyomma echidnae*. The names of these parasites indicate some degree of host specificity, but again there are simply not enough data to be certain if that is the case or to determine the significance of individual arthropod parasites. Ticks and fleas can be observed on almost every echidna encountered in the field. Very often the spherical grey bodies of female ticks can be seen between the spines and especially in the ear slits. The latter site is probably safe from the grooming activities of the host, although echidnas have amazing agility when it comes to using the long grooming claws on the hindfeet. Aided by the degree to which the hindfoot can swivel, echidnas are quite capable of scratching between the spines on the back of their neck as well as elsewhere.

Tapeworms are sometimes seen as a white, squirming item hanging out of the cloaca of an echidna or on a fresh dropping. These are likely to be *Echidnotaenia tachyglossi* or *Linstowia echidnae*. Both of these tapeworms seem to be common in echidnas, but the degree to which they debilitate the host is unknown.

Captive housing

Both *Tachyglossus* and *Zaglossus* are relatively easy animals to maintain in captivity although some elementary precautions must be taken. Both are excellent and determined diggers and a concrete or chain link barrier must be extended at least a metre below the surface at the boundary of the enclosure. A chain link mesh buried beneath the soil is useful, but as soon as even one or two links begin to deteriorate a determined echidna can break through.

Echidnas are also proficient at climbing chain link fences and can be over the top of a two-metre-high fence without difficulty. However, they can injure themselves if they fall in the process, so if such fencing is used there should be a barrier preventing toeholds at the bottom.

The floor of an enclosure is important as echidna feet become cracked and infected on hard surfaces such as concrete or metal, and the problem is made worse by covering such surfaces with sawdust. It is likely that the loss of newly hatched echidnas in captivity has been due to lack of natural substrate. The young is lost from the rudimentary pouch and not taken back up by the female. All echidna enclosures for long-term captivity should include an area with natural soil substrate.

Echidnas present zookeepers with a common problem because if they supply a complex environment with logs to climb in and dirt to dig in, the echidnas will probably love it but the zoo visitors will never see them. If there is a spot of morning sun to bask in at the very back of the enclosure, that is where the echidnas will congregate. It is a good display technique to provide a clear area for basking in the morning or winter sun that is visible to zoo visitors. It is always a good idea to have the feeding point in view of visitors, as echidnas are fearless when it comes to feeding time in captivity. Display of *Zaglossus* in nocturnal houses has worked well in several Australian exhibits, but *Tachyglossus* is more inclined to avoid visitors, especially the noise they produce, regardless of diurnal cycles. Even in the wild their activity cycle is variable, depending upon temperature.

Temporary housing of an echidna, such as might be required to transport an injured animal to a place of care or to transport a recovered animal to a point of release, must be chosen with care. Even an apparently disabled echidna can demolish a cardboard box in seconds. Any place of purchase, such as a space between the planks of a wooden box for their claws, can provide a point of leverage. The timbers are split and broken away facilitating escape.

Echidnas are even tolerant of self-injury as demonstrated by the loss of skin on the legs when they have broken out of cords tied around their legs. Wire cages should not be used to transport echidnas as they can damage their snout in trying to escape. Skin on the snout torn by sharp edges will bleed profusely. An echidna wedged under a car seat can only be removed by disassembly of the surrounding automobile! For short periods a large plastic garbage bin is an ideal carrier, provided it cannot be tipped over and the echidna cannot get its snout over the rim of the container. If the echidna can get its snout over the rim of the container, it will use its strong neck muscles to lever its body up until its front claws are over the rim – and it is off!

On hot days, care must be taken to keep the animal from overheating. We have found that an echidna will remain calm and quiet for long periods if something as simple as a burlap bag is supplied for it to take cover underneath.

Captive diet

Several inventive diets have been devised by zookeepers for echidnas. Those containing milk are, at best, wasteful, since echidnas lack the digestive enzyme lactase. At worst, milk will cause diarrhoea, weight loss and eventual death.

Milk is not a natural food for any adult animal and modern zoo diets usually avoid it. It is impractical to provide echidnas with their natural diet, and they tend to ignore the adult workers of almost any ant species readily available.

The prize for the most unnatural diet ever used for echidnas must go to the one in use by Taronga Zoo in 1973. It consisted of:

1 tablespoon	minced meat
1 tablespoon	minced carrot
57 grams	horse blood
½	boiled egg and shell
1 tablespoon	Lactogen
1 tablespoon	condensed milk
½ teaspoon	clay
1	lettuce leaf
1 drop	formic acid
100 ml	water

Formic acid might seem reasonable for an anteater, but feeding trials have shown that it does not increase the palatability of food for echidnas nor does it have any effect on their health. The inclusion of carrot for an anteater is a little harder to understand, but it is the lettuce leaf that raises this from the silly to the absurd. Of course the Lactogen and condensed milk lead to diarrhoea, but presumably the clay slowed that down a bit.

The inclusion of inert bulk in the diet is a very good and a very natural thing. Echidna scats (droppings) in the wild are cylindrical casts of dirt, sand, grit and insect exoskeletons. Echidnas take in a great deal of dirt with their sticky tongue while feeding. We have at times added a large percentage of autoclaved soil to captive echidna diets and found that it was readily eaten and there was no diarrhoea. Because of the preparation time involved, we later replaced the soil with pollard, which is the fibrous and relatively inert outer covering of the wheat grain. The diet that we found to be easily prepared and readily eaten was as follows:

5 parts	pollard	
4 parts	high quality meat meal or minced, cooked lean beef	
1 part	Glucodin (a glucose powder)	
1 drop	pentavite (a multivitamin preparation)	
water	as required to make a porridge-like consistency	

We have successfully kept echidnas on this diet from a few days to five years. Most individuals took to it readily, but regardless of the diet there are always a few echidnas that will not take a captive diet for up to several weeks after initial capture. Since a fat echidna can go for a couple of months without feeding, it is simply a matter of waiting until the animal is ready. In many cases the problem is quite the opposite; if too much food is available it will eventually have to be restricted as the echidnas come to resemble spiny butterballs.

Presentation of food requires some inventiveness since the inevitable response of echidnas to food presented in a standard dish is either to sit in it or turn it over repeatedly. A heavy trough with a cover with holes of about five centimetres diameter works well.

Finally, the success of a diet may depend on individual echidnas as well as the skill of the keeper. The record life span of 50 years reported for an echidna at the Philadelphia Zoo was on a diet of half a pint of whole milk and a raw egg daily!

Longevity in zoos

The echidna made its zoo debut in 1845 at the Zoological Gardens of London. It lived only four days. Much greater success has been obtained since then by European and American zoos, as well as Australian zoos.

The longevity of some echidnas held in zoos outside Australia is as follows:

Zaglossus bruijni
Berlin	1906(1911?)–1943	31(36?) years
London	1912–1943	30 years 8 months
	1912–1917	6 years 3 months
Amsterdam	1911–1920	8 years 8 months

Tachyglossus aculeatus
    Philadelphia   1903–1953   50 years

From data contained in the International Zoo Yearbook, the following zoos are the only ones in which captive-bred echidnas have hatched:

Berlin	1908
Basel	1955, 1967
Sydney (Taronga)	1977
Philadelphia	1982, 1986
Oklahoma City	1988

At the Oklahoma Zoo, mating occurred in December–January and the young was first observed in the pouch on 17 March. It was found out of the pouch 58 days after the calculated time of hatching and at that time weighed 128 g. Clearly the reproductive cycle had shifted six months to follow the northern hemisphere seasons.

None of the captive-bred echidnas hatched in captivity has survived to maturity. There is clearly little role for zoos in the preservation of echidnas, and conservation of the rarest and most endangered monotreme, *Zaglossus bruijni*, is a matter of direct and immediate action in New Guinea.

Glossary

ambient temperature	the temperature of an organism's surroundings; the sum of air temperature, radiant heat, conducted heat, etc. – 'environmental temperature' is often used as a synonym
ancestral (character)	a character present in a taxon that is also present in its immediate ancestor.
anlage	the undifferentiated embryonic precursor of a structure.
apomorphic	a term used by evolutionary biologists to refer to characters possessed by a taxon by virtue of their having evolved within that taxon. (Also known as 'derived' characters. Fur, for example, is an apomorphic character for the taxon Mammalia.)
archaic	ancient. When used in an evolutionary sense, the term usually refers to characters that appeared very early in a fossil lineage or in ancestors to that lineage.
areolar patches	areas on the skin where milk ducts open. An odd term, since 'areolae' is derived from Latin meaning 'small patches'.
asphyxia	a dangerous situation in which blood levels of oxygen drop and carbon dioxide rise in response to environmental conditions. Encountered by diving and burrowing animals.
benthic	literally 'of the deep' – refers to aquatic organisms living at or near the bottom as opposed to those that live at or near the surface (pelagic).
biological species	all organisms that have the potential to exchange genetic information (interbreed).
chitin	a tough polysaccharide forming, along with some protein, the exoskeleton of insects.
cline	gradual change in gene frequencies or character states within a species, across its geographical distribution.

cloaca	the chamber into which the ureters, large intestine, female and male reproductive tracts enter. The cloaca then opens through a single opening to the outside. That opening itself has come to be called (incorrectly) the cloaca.
convergent	literally 'coming together'. Convergent characters are similar characters (usually but not necessarily morphological) adapted to similar ways of life in taxa that are unrelated, except through distant ancestors that lacked those characters. Bat wings vs. bird wings are an example.
cornified	tissue that is hardened due to the presence of keratin.
crepuscular	active at dawn and dusk, although most crepuscular mammals have a greater activity at dusk than in the morning.
cretaceous	the last period of the Mesozoic era. Terminated with the extinction of the dinosaurs about 65 million years ago.
cribriform plate	a sieve-like bony plate at the base of the skull through which a number of small nerves pass.
crural	pertaining to the hind limb.
derived (character)	see 'apomorphic'.
desmosomes	specialised cell junctions between adjacent cell membranes.
diurnal	active during daylight hours.
edentate	(noun) the state of being without teeth. It is also a taxon of placental mammals including a number of anteaters.
edentulous	lacking teeth.
endangered	IUCN definition.
end organ	specialised ending of a sensory nerve.
endothermy/ endothermic	endothermy (noun) is the metabolic state of an organism in which internal body processes can maintain internal body temperatures above those available in the environment. Only birds and

	mammals are endothermic. 'Warm blooded' was used in older literature as a synonym.
epidermis	outer, keratinised layer of the skin.
euthermia/euthermic	the normal body temperature range of an individual. Used most commonly to refer to the body temperature range of hibernators (such as echidnas) when they are active as opposed to the much lower values possible during hibernation.
hibernaculum(a)	a place chosen or constructed by a mammal or bird as the site in which they will hibernate over winter.
instar	a stage in the larval development of an insect between two moults.
keratin	(noun) protein toughened by strong sulfide bonds between protein molecules.
keratinised	tissues containing keratin that are therefore hardened. Fingernails and hair are keratinised tissues.
keratinocyte	cell in epidermis that produces keratin.
keratinous	same as 'keratinised'.
lamellated	layered like an onion.
lepidopteran	the taxon of insects that contains moths and butterflies.
lingual	related to the tongue. In the case of teeth, lingual refers to the surface next to the tongue as opposed to labial which is next to the lips.
lumen	the hollow part of a tube. The blood flows through the lumen of veins.
manus	the hand including the metacarpals, carpals and phalanges and all the tissues thereon.
mechanoreceptor	a receptor that responds to a physical stimulus such as pressure
merkel cells	sensory cells with light touch nerve endings.
mesic	an environment that is neither extremely dry nor extremely wet.
mesozoic	the geological era from about 570 to 65 million years ago; includes the Triassic, Jurassic and Cretaceous periods.

monotreme	an egg-laying mammal. The name refers to the fact that there is only one (mono) opening ('treme' from the Greek) to the outside of the body through which waste and reproductive products pass.
morphological	(adjective) structural or anatomical.
morphology	(noun) overall structure. The morphology of an animal includes its size, shape and texture as well as the individual skeletal elements and organs that comprise it. The term 'anatomy' usually implies only the latter.
mosaic	a structure composed of different elements, as for example a picture made up of many different coloured tiles.
myelinated	a nerve with a surrounding layer composed mainly of myelin.
myrmecophagy(ges)	refers to a diet comprised mainly of ants and termites. It is a specialised type of insectivory.
niche	the functional position of an organism in a community, together with its environmental position with respect to factors such as temperature, rainfall, etc.
nocturnal	active during hours of darkness.
oviparity	egg-laying; as opposed to viviparity which is giving live birth.
pacinian corpuscle	deep pressure sensory nerve endings.
parasaggital	parallel with the median or mid-line plane. Used most commonly to describe a section of an organ, e.g. the brain, that does not divide it into two equal left and right halves but is parallel to the section that does.
pelage	similar to the word 'pelt'. The outer coat of fur or its derivatives such as hair, bristles and spines.
pentadactylous	having a hand or foot with five digits.
peristalsis	involuntary muscular contractions of the gut wall that move the contents along.
pes	the hind foot including the tarsals, metatarsals and phalanges and covering tissues.

pheromone	a substance secreted to the outside of the body by an animal and which influences the behaviour of another individual; usually associated with reproduction.
pinna	the external ear. Usually a flap shaped structure (although the Latin word means 'feather') that helps direct sound into the ear drum.
pleistocene	the epoch that preceded the Holocene (Recent). From about 2.2 million to 10 000 years ago.
plesiomorphic	a term used for characters that are shared by different lineages of animals and have been inherited from a common ancestor. The word literally means 'old-featured'. This term is preferred in modern evolutionary to the term 'primitive' (see next entry).
primitive	a word used to describe taxa or characters that are considered to represent an early level of evolution compared to some other character or taxon. Because the word also implies inefficiency or an inferior state, the word is not used by evolutionary biologists. The terms 'plesiomorphic' or 'ancestral' are usually more appropriate.
prototheria	a mammalian taxon including monotremes and some extinct forms.
pyloric	the posterior part of the stomach just before the small intestine. Has much thicker walls and is glandular relative to the anterior (cardiac) portion.
quaternary	the second period of the Cenozoic era, including the Pleistocene and Holocene (or Recent) epochs.
radiation	in biological terms a radiation occurs when a number of specialised forms evolve from a generalised ancestor. Living Australian marsupials represent a radiation from a small, carnivorous ancestor.
ruffini ending	touch-sensitive nerve ending.
scats	a polite word for shit (faecal droppings).
secondary palate	that part of the palate that is cartilaginous as opposed to the bony anterior portion; also known as the soft palate.

spatulate	shaped like a spade or shovel.
spheroids	horizontally flattened dilations of the vesicle chain receptor axon.
taxon (pl. taxa)	a formal group of organisms. May be large (Class) or small (Genus). Ideally a taxon should include all the descendants of a common ancestor.
tertiary	the first period part of the Cenozoic Era, including the Paleocene, Eocene, Oligocene, Miocene and Pliocene epochs. See also 'Quaternary'.
testicond	the state wherein the testes remain inside the body and do not descend into a scrotum.
thalamus	part of the brain underneath the cerebral cortex that functions mainly as a switching area, connecting and distributing various nervous pathways.
therapsid	a diverse taxon traditionally classified as 'reptiles' but having evolved separately from dinosaurs, lizards, snakes , turtles, etc. Therapsids were the dominant land vertebrates of the early Mesozoic and many characters found only among mammals today appeared in terminal therapsid lineages.
therian	the mammalian clade including marsupials and placentals but excluding monotremes.
trigeminal nerve	a cranial nerve that carries, amongst other things, sensory supply from the face and nasal cavities to the brain.
tympanic	referring to the ear.
vacuole	membrane-bound sac found within a cell.
vestigial	something that is present but has no discernable function. The assumption is that there was a function in the past, but while that function has ceased the organ or phenomenon has remained. The spur on the male echidna is a good example. It is there but it does not do anything.

Bibliography

(Note: items marked * are good general resources)

*Augee, M.L. (ed.) (1978). *Monotreme Biology*. The Royal Zoological Society of NSW, Sydney.

*Augee. M.L. (ed.) (1992). *Platypus and Echidnas*. The Royal Zoological Society of NSW, Sydney.

*Augee, M.L. (ed.) (2004). Monotreme papers. In *Proceedings of the Linnean Society of New South Wales*, Volume 125, Sydney.

Beard, L.A., Grigg, G.C. and Augee, M.L. (1992). Reproduction by echidnas in a cold climate. In *Platypus and Echidnas* (ed. M.L. Augee) pp. 93–100. The Royal Zoological Society of NSW, Sydney.

Bentley, P.J. and Schmidt-Nielsen, K. (1967). Role of the kidney in water balance of the echidna. *Comparative Biochemistry and Physiology* 20, 285–290.

Buchmann, O.L. and Rhodes, J. (1978). Instrumental learning in the echidna *Tachyglossus aculeatus setosus*. In *Monotreme Biology* (ed. M.L. Augee) pp. 131–145. The Royal Zoological Society of NSW, Sydney.

Collins, L.R. (1973). *Monotremes and Marsupials, a Reference for Zoological Institutions*. Smithsonian Institute Press, Washington DC.

Dobroruka, Ludek J. (1960). Einige Beobachtungen an Ameisenigeln, *Echidna aculeata* Shaw (1792) [Some observations on the 'Ant-Hedgehog', *Echidna aculeata* Shaw (1792)]. *Zeitschriftfur Tierpsychologie* Band 17, Heft 2, 178–181.

Flannery, T.F. and Groves, C.P. (1998). A revision of the genus *Zaglossus* (Monotremata, Tachyglossidae), with description of a new species and subspecies. *Mammalia* 62, 367–396.

Gates, G.R. (1978). Vision in the monotreme anteater (*Tachyglossus aculeatus*). In *Monotreme Biology* (ed. M.L. Augee) pp. 147–169. The Royal Zoological Society of NSW, Sydney.

*Grant, T.R. (2002). *The Platypus*. UNSW Press, Sydney.

*Griffiths, M. (1968). *Echidnas*. Pergamon Press, London.

*Griffiths, M. (1978). *Biology of the Monotremes*. Academic Press: New York.

*Griffiths, M. (1989). Tachyglossidae. In *Fauna of Australia* Vol. 1B Mammalia (eds D.W. Walton and B.J. Richardson) pp. 407–435. Australian Government Publishing Service, Canberra.

Grigg, G.C., Beard, L. and Augee, M.L. (1992). Thermal relations of free-living echidnas during activity and in hibernation in a cold climate. In *Platypus and Echidnas* (ed. M.L. Augee) pp. 160–163. The Royal Zoological Society of NSW, Sydney.

Grigg, G.C., Beard, L.A. and Augee, M.L. (2004). The evolution of endothermy and its diversity in mammals and birds. *Physiological and Biochemical Zoology* 77, 982–997. Kermack, K.A., Mussett, F. and Rigney, H.W. (1973). The lower jaw of *Morganucodon*. *Zoological Journal of the Linnean Society* 53, 87–175.

Martin, C.J. (1902). Thermal adjustment and expiratory exchange in monotremes and marsupials. *Philosophical Transactions of the Royal Society London* series B, **195**, 1–37.

McNab, B.K. (1984). Physiological convergence amongst ant-eating and termite-eating mammals. *Journal of Zoology London* **203**, 485–510.

Murray, P.F. (1984). Furry egg-layers: The monotreme radiation. In *Vertebrate Zoogeography and Evolution in Australasia* (eds M. Archer and G. Clayton) pp. 571–583. Hesperian Press, Perth.

Musser, A.M. (2003). A review of the monotreme fossil record and comparison of palaeontological and molecular data. *Comparative Biochemistry and Physiology Part A* **136**, 927–942.

Proske, U. (1990). Electric monotreme. *Natural History* **23**, 288–295.

Rich, T.H., Hopson, J.A., Musser, A.M., Flannery, T.F. and Vickers-Rich, P. (2005). Independent origins of middle ear bones in monotremes and therians. *Science* **307**, 910–914.

*Rismiller, P.D. (1999). *The Echidna. Australia's Enigma*. Hugh Lauter Levin Associates, Inc., Publishers Group West.

Rismiller, P.D. and McKelvey, M.W. (2000). Frequency of breeding and recruitment in the short-beaked echidna, *Tachyglossus aculeatus*. *Journal of Mammalogy* **81**, 1–17.

Romer, A.S. (1956). *The Osteology of Reptiles*. University of Chicago Press, Chicago.

Schmidt-Nielsen, K. (1972). *How Animals Work*. Cambridge University Press, Cambridge.

*Semon, R. (1899). *In the Australian Bush*. Macmillan & Co., London.

Smith, G.E. (1902). Mammalia. Order Monotremata. In *The Royal College of Surgeons Museum Physiological Catalogue*, Vol. 2, pp. 138–157. Taylor and Francis, London.

Whittington, R.J. (1992). Role of infectious diseases in the population biology of monotremes. In *Platypus and Echidnas* (ed. M.L. Augee) pp. 285–293. The Royal Zoological Society of NSW, Sydney.

Index

References in bold refer to illustrations

activity patterns   90–1
albino   2, 8, **26**
anteaters   *see* convergence
beak   *see* snout
bill   *see* snout
brain   45–9, **46, 48**
burrows   60, 65, 93, 111, 116, 118
   nursery   **31**, 78, 82, 84, 86–8
   platypus   11
captive management   122–25
classification   *see* taxonomy
claws   3, 8, 9, **25, 28**, 42–3, 85
conservation   119–20
convergence   20–1, **21**, 113
courtship   **29**, 77–80
defence   20–1, 89, 92–3, 119
diet, captivity   124–25
   natural   *see* food
digestion   106
digging   99–103, **101**
discovery, by Europeans   4–5
disease   121–22
distribution   6, 9, 119
drinking   **32**, 115–16
ear   39, 59–65, **63**
   cochlea   39, 62–4, **64**
   ossicles   39, **61**
   pinna   2, 22, 59–60, **60**
Echidna hystrix   5
egg   82, 84
egg tooth   37–8, **37**
electroreceptors   11, 68–70, **68, 69**
endocrines   116–17
eyes   2
food   18, 97–8
foraging   98–100
fossil forms   10, 17–20

gestation   82, 84
grooming   3, **27**, 122
habitat   6–7, 120
hearing   59–65
heart rate   52, 108, 109, 116
hibernation   111–17
home range   91–2
hunting, of   119
incubation   78, 84, 110, 111
intelligence   49, 50–1
learning   50–1
life expectancy   *see* longevity
limbs   2, 3, 42–3
long-beaked echidna   7–10, 111
longevity   7, 125–26
mammary glands   4, 85–6, **85**
mastication   105–06
mating   **29**, 77–9, 80–2, 92, 126
Megalibgwilia   18
Mesozoic monotremes   13
metabolic rate   113–14
metabolism   93
milk   86–7, 124
nest   89–90, 92
nose   *see* snout
Ornithorhynchus   *see* platypus
oviparity   1, 22, 77
parasites   122
pectoral girdle   *see* shoulder
penis   **81**
pheromone   79
platypus   10–12
pouch   4, 8, **30**, 84, 85
pouch young   **30**, **37**, 85, 126
predators   120
prey   95–7, **96**
primitive characters   22–3, 36, 39, 40, 41, 93
push rods   70–4, **71**

quills *see* spines
relationships 12, 16
reproductive cycle 78
reproductive tracts
  female 83
  male 79, 80
respiration 52, 108–09, 118
scats 29, 98, 106, 124
scent trails 79
shoulder 40–2, **41**, 101
skeleton 35–43
skull 36–9
sleep 51–3
smell, sense of 66–7, 86, 98
snout 26, 27, 33–5, 65–6, **66**, 65–75, 68, 103–04
spinal cord 53–4

spines 6, 8, 35
spur 3–4, **3**, 8, 11, 43, 92
Steropodon 14
subspecies 5–7, 8, 9
swimming **32**, 92, 114, 116
taxonomy 8–9, 12
Teinolophos 14
temperature, body 112–14
territory *see* home range
tongue 8, 10, 47, 48, 95, 97, 99, 104–05, **104**, **106**
vertebrae 40
vision 51, 55–9, 90, 99
water balance 114–16
weaning 86–8
Zaglossus *see* long-beaked echidna
zoo records 125–26